Learn Business Analytics in Six Steps Using SAS and R

A Practical, Step-by-Step Guide to Learning Business Analytics

Subhashini Sharma Tripathi

Apress®

Learn Business Analytics in Six Steps Using SAS and R

Subhashini Sharma Tripathi
Bangalore, Karnataka
India

ISBN-13 (pbk): 978-1-4842-1002-4   ISBN-13 (electronic): 978-1-4842-1001-7
DOI 10.1007/978-1-4842-1001-7

Library of Congress Control Number: 2016961720

Managing Director: Welmoed Spahr
Lead Editor:Celestin Suresh John
Technical Reviewer: Ujjwal Dalmia
Editorial Board: Steve Anglin, Pramila Balan, Laura Berendson, Aaron Black, Louise Corrigan,
    Jonathan Gennick, Robert Hutchinson, Celestin Suresh John, Nikhil Karkal, James Markham,
    Susan McDermott, Matthew Moodie, Natalie Pao, Gwenan Spearing
Coordinating Editor: Prachi Mehta
Copy Editor: Kim Wimpsett
Compositor: SPi Global
Indexer: SPi Global
Artist: SPi Global

Distributed to the book trade worldwide by Springer Science+Business Media New York,
233 Spring Street, 6th Floor, New York, NY 10013. Phone 1-800-SPRINGER, fax (201) 348-4505, e-mail orders-ny@springer-sbm.com, or visit www.springeronline.com. Apress Media, LLC is a California LLC and the sole member (owner) is Springer Science + Business Media Finance Inc (SSBM Finance Inc). SSBM Finance Inc is a **Delaware** corporation.

For information on translations, please e-mail rights@apress.com, or visit www.apress.com.

Apress and friends of ED books may be purchased in bulk for academic, corporate, or promotional use. eBook versions and licenses are also available for most titles. For more information, reference our Special Bulk Sales–eBook Licensing web page at www.apress.com/bulk-sales.

Any source code or other supplementary materials referenced by the author in this text are available to readers at www.apress.com. For detailed information about how to locate your book's source code, go to www.apress.com/source-code/. Readers can also access source code at SpringerLink in the Supplementary Material section for each chapter.

Printed on acid-free paper

Contents at a Glance

Contents

About the Author

Subhashini Sharma Tripathi is an analytics enthusiast. After working for a decade with GE Money, Standard Chartered Bank, Tata Motors Finance, and Citi GDM, she started teaching, blogging, and consulting in 2012. As she worked, she became convinced that analytics and data science help reduce dependency on experience. Further, she believes it gives modern managers a conclusive way to solve many real-world problems faster and more accurately. In this evolving business landscape, it also helps define longer-term strategies and makes better choices available. In other words, you can get "more bang for your buck" with analytics.

Subhashini is the founder of pexitics.com, and her first product is the Pexitics Talent Score, a pre-interview score. The company makes tools for effective human resource management and consults in analytics.

You can connect with her via LinkedIn at https://in.linkedin.com/in/subhashinitripathi or via e-mail with subhashini@pexitics.com.

Acknowledgments

This book is my first, and the experience of writing it has been an exciting and bumpy journey. This book and its writing coincided with the creation and launch of pexitics.com.

The journey would not have been possible without a lot of support and encouragement from my family and the editorial team at Apress, especially Celestin Suresh John, for ensuring that my morale did not flag on the way. I express my heartfelt gratitude to my mother, Dr. M. Tripathi (PhD), for her support and help in words, deeds and prayers.

My thought process has been significantly influenced by the book *Basic Business Statistics* (12th edition) by Mark L. Berenson, David M. Levine, and Timothy C. Krehbiel. I read about the DCOVA process in that book. As I worked with that process, I added another stage, called Insight Generation, and now use the process of DCOVA and I.

When I started my journey into number-based decision-making in 2002, there was a dearth of structured mentoring, and a lot of things were self-discovered and self-taught. I have written this book so that analytics and data science aspirants can start on the journey in a structured way and with a lot of confidence to solve real business problems.

The next edition will cover predictive models.

Introduction

In the last decade, analytics and data science have come into the forefront as support functions for business decisions. A decade ago, business analytics was a little-known career choice. With the drastic dip in data storage costs and the huge increase in data volumes (projected to hit 40 zettabytes in 2020), chief experience officers (CXOs) and modern managers now need analytics and data science to make informed decisions at every point.

Have you wondered how to get started on a career in analytics and data science?

This book teaches you how to solve problems and execute projects in analytics through the Define, Collect, Organize, Visualize, Analyze, and Insights (DCOVA and I) process. Thus, even when the data is very new or the problem is not familiar, you can solve it by using a step-by-step checklist for deduction and inferencing. Finally, for implementing analytics output, the conclusion or insight needs to be understood in plain business terms.

This book teaches you how to do analytics on business data using two popular software tools, SAS and R. SAS is licensed software that is the leader in the sectors that have regulatory supervision (banking, clinical research, insurance, and so on). R is open source software that is popular in sectors without regulators such as retail, technology (including ITES), BPOs, and so on. So, irrespective of the industry in which you work, this book will provide you with the knowledge and skills you and your managers need to make better decisions faster.

You no longer need to choose between the two most popular software tools.

How can business turn this data into useful information in a reasonably fast turnaround time?

This question becomes important for running a successful business. Only if the information is available to management at the correct time will the business be able to make the correct decisions. For this, you need business analytics, loosely described as doing statistics on large volumes of data, to arrive at conclusions and models that will aid business decision-making.

The statistical techniques can be divided into the five broad segments of descriptive statistics, inferential statistics, differences statistics, associative statistics, and predictive statistics. I will cover models related to associative and predictive stats in the next edition. In this book, I will focus on developing your understanding of the process of problem-solving and the statistics related to the descriptive, differences, and associative statistical techniques.

Do connect with me via LinkedIn at `https://in.linkedin.com/in/subhashinitripathi` or via e-mail with `subhashini@pexitics.com`.

CHAPTER 1

■ ■ ■

The Process of Analytics

In this chapter, you will look at the process and evolution of analytics. These are some of the topics covered:

- The process of analytics
- What analytics is
- The evolution of analytics
- The dawn of business intelligence

What Is Analytics? What Does a Data Analyst Do?

A casual search on the Internet for *data scientist* offers up the fact that there is a substantial shortage of manpower for this job. In addition, Harvard Business Review has published an article called "Data Scientist: The Sexiest Job of the 21st Century" (`http://hbr.org/2012/10/data-scientist-the-sexiest-job-of-the-21st-century/ar/1`). So, what does a data analyst actually do?

To put it simply, *analytics* is the use of numbers or business data to find solutions for business problems. Thus, a *data analyst* looks at the data that has been collected across huge enterprise resource planning (ERP) systems, Internet sites, and mobile applications.

In the "old days," we just called upon an expert, who was someone with a lot of experience. We would then take that person's advice and decide on the solution. It's much like we visit the doctor today, who is a subject-matter expert.

As the complexity of business systems went up and we entered an era of continuous change, people found it hard to deal with such complex systems that had never existed before. The human brain is much better at working with fewer variables than many. Also, people started using computers, which are relatively better and unbiased when it comes to new forms and large volumes of data.

An Example

The next question often is, what do I mean by "use of numbers"? Will you have do math again?

The last decade has seen the advent of software as a service (SaaS) in all walks of information gathering and manipulation. Thus, analytics systems now are button-driven systems that do the calculations and provide the results. An analyst or data scientist has to look at these results and make recommendations for the business to implement. For example, say a bank wants to sell loans in the market. It has data of all the customers who have taken loans from the bank over the last 20 years. The portfolio is of, say, 1 million loans. Using this data, the bank wants to understand which customers it should give pre-approved loan offers to.

Electronic supplementary material The online version of this chapter (doi: 10.1007/978-1-4842-1001-7_1) contains supplementary material, which is available to authorized users.

The simplest answer may be as follows: all the customers who paid on time every time in their earlier loans should get a pre-approved loan offer. Let's call this set of customers Segment A. But on analysis, you may find that customers who defaulted but paid the loan after the default actually made more money for the bank because they paid interest plus the late payment charges. Let's call this set Segment B.

Hence, you can now say that you want to send out an offer letter to Segment A + Segment B.

However, within Segment B there was a set of customers who you had to send collections teams to their homes to collect the money. So, they paid interest plus the late payment charges minus the collection cost. This set is Segment C.

So, you may then decide to target Segment A + Segment B – Segment C.

You could do this exercise using the decision tree technique that cuts your data into segments (Figure 1-1).

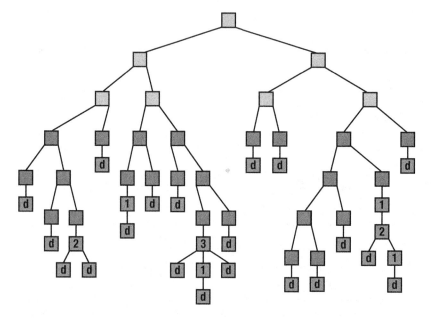

Figure 1-1. *Decision tree*

A Typical Day

The last question to tackle is, what does the workday of an analytics professional look like? It probably encompasses the following:

- The data analyst will walk into the office and be told about the problem that the business needs input on.

- The data analyst will determine the best way to solve the problem.

- The data analyst will then gather the relevant data from the large data sets stored in the server.

- Next, the data analyst will import the data into the analytics software.

- The data analyst will run the technique through the software (SAS, R, SPSS, XLSTAT, and so on).

- The software will produce the relevant output.

- The data analyst will study the output and prepare a report with recommendations.
- The report will be discussed with the business.

Is Analytics for You?

So, is analytics the right career for you? Here are some points that will help you decide:

- *Do you believe that data should be the basis of all decisions?* Take up analytics only if your answer to this question is an unequivocal yes. Analytics is the process of using and analyzing a large quantum of data (numbers, text, images, and so on) by aggregating, visualizing/creating dashboards, checking repetitive trends, and creating models on which decisions can be made. Only people who innately believe in the power of data will excel in this field. If some prediction/analysis is wrong, the attitude of a good analyst is that it is because the data was not appropriate for the analysis or the technique used was incorrect. You will never doubt that a correct decision will be made if the relevant data and appropriate techniques are used.

- *Do you like to constantly learn new stuff?* Take up analytics only if your answer to this question is an unequivocal yes. Analytics is a new field. There is a constant increase in the avenues of data currently regarding Internet data, social networking information, mobile transaction data, and near field communication devices. There are constant changes in technology to store, process, and analyze this data. Hadoop, Google updates, and so on, have become increasingly important. Cloud computing and data management are common now. Economic cycles have shortened, and model building has become more frequent as older models get redundant. Even the humble Excel has an Analysis ToolPak in Excel 2010 with statistical functions. In other words, be ready for change.

- *Do you like to interpret outcomes and then track them to see whether your recommendations were right?* Take up analytics only if your answer to this question is an unequivocal yes. A data analyst will work on a project, and the implementation of the recommendations will generally be valid for a reasonably long period of time, perhaps a year or even three to five years. A good analyst should be interested to know how accurate the recommendations have been and should want to track the performance periodically. You should ideally also be the first person to be able to say when the analysis is not working and needs to be reworked.

- *Are you ready to go back to a text book and brush up on the concepts of math and statistics?* Take up analytics only if your answer to this question is an unequivocal yes. To accurately handle data and interpret results, you will need to brush up on the concepts of math and statistics. It becomes important to justify why you chose a particular path during analysis versus others. Business users will not accept your word blindly.

- *Do you like debating and logical thinking?* Take up analytics only if your answer to this question is an unequivocal yes. As there is no one solution to all problems, an analyst has to choose the best way to handle the project/problem at hand. The analyst has to be able to not only know the best way to analyze the data but also give the best recommendation in the given time constraints and budget constraints. This sector generally has a very open culture where the analyst working on a project/problem will be required to give input irrespective of the analyst's position in the hierarchy.

Do check your answers to the previous questions. If you said yes for three out of these five questions and an OK for two, then analytics is a viable career option for you. Welcome to the world of analytics!

Evolution of Analytics: How Did Analytics Start?

As per the Oxford Dictionary, the definition of *statistics* is as follows:

> *The practice or science of collecting and analyzing numerical data in large quantities, especially for the purpose of inferring proportions in a whole from those in a representative sample.*[1]

Most people start working with numbers, counting, and math by the time we are five years old. Math includes addition, subtraction, theorems, rules, and so on. Statistics is when we start using math concepts to work on real-life data.

Statistics is derived from the Latin word *status*, the Italian word *statista*, or the German word *statistik*, each of which means a political state. This word came into being somewhere around 1780 to 1790.

In ancient times, the government collected the information regarding the population, property, and wealth of the country. This enabled the government to get an idea of the manpower of the country and became the basis for introducing taxes and levies. Statistics are the practical part of math.

The implementation of standards in industry and commerce became important with the onset of the Industrial Revolution, where there arose a need for high-precision machine tools and interchangeable parts. *Standardization* is the process of developing and implementing technical standards. It helps in maximizing compatibility, interoperability, safety, repeatability, and quality.

Nuts and bolts held the industrialization process together; in 1800, Henry Maudslay developed the first practical screw-cutting lathe. This allowed for the standardization of screw thread sizes and paved the way for the practical application of interchangeability for nuts and bolts. Before this, screw threads were usually made by chipping and filing manually.

Maudslay standardized the screw threads used in his workshop and produced sets of nuts and bolts to those standards so that any bolt of the appropriate size would fit any nut of the same size.

Joseph Whitworth's screw thread measurements were adopted as the first unofficial national standard by companies in Britain in 1841 and came to be known as the British standard Whitworth.

By the end of the 19th century, differences and standards between companies were making trading increasingly difficult. The Engineering Standards Committee was established in London in 1901 and by the mid-to-late 19th century, efforts were being made to standardize electrical measurements. Many companies had entered the market in the 1890s, and all chose their own settings for voltage, frequency, current, and even the symbols used in circuit diagrams, making standardization necessary for electrical measurements.

The International Federation of the National Standardizing Associations was founded in 1926 to enhance international cooperation for all technical standards and certifications.

The Quality Movement

Once manufacturing became an established industry, the emphasis shifted to minimizing waste and therefore cost. This movement was led by engineers who were, by training, adept at using math. This movement was called the *quality movement*. Some practices that came from this movement are Six Sigma and just-in-time manufacturing in supply chain management. The point is that all this started in the Industrial Revolution in 1800s.

This was followed with the factory system with its emphasis on product inspection.

---

[1]www.oxforddictionaries.com/definition/english/statistics

After the United States entered World War II, the quality became a critical component since bullets from one state had to work with guns manufactured in another state. For example, the U.S. Army had to inspect manually every piece of machinery, but this was very time-consuming. Statistical techniques such as sampling started being used to speed up the processes.

Japan around this time was also becoming conscious of quality.

The quality initiative started with a focus on defects and products and then moved on to look at the process used for creating these products. Companies invested in training their workforce on Total Quality Management (TQM) and statistical techniques.

This phase saw the emergence of seven "basic tools" of quality.

- Cause-and-effect diagram

- Check sheet

- Control charts

- Histogram

- Pareto chart

- Scatter diagram

- Stratification/flowchart/run chart

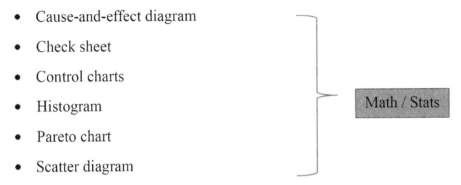

Statistical Process Control from the early 1920s is a method of quality control using statistical methods, where monitoring and controlling the process ensures that it operates at its full potential. At its full potential, a process can churn out as much conforming product or standardize a product as much as possible with a minimum of waste.

This is used extensively in manufacturing lines with a focus on continuous improvement and is practiced in these two phases:

- Initial establishment of the process

- Regular production use of the process

The advantage of Statistical Process Control (SPC) over the methods of quality control such as inspection is that it emphasizes early detection and prevention of problems rather than correcting problems after they occur.

The following were the next steps:

- *Six Sigma*: A process of measurement and improvement perfected by GE and adopted by the world

- *Kaizen*: A Japanese term for continuous improvement; a step-by-step improvement of business processes

- *PDCA*: Plan-Do-Check-Act, as defined by Deming

What was happening on the government front? The maximum data was being captured and used by the military. A lot of the business terminologies and processes used today have been copied from the military: sales *campaigns*, marketing *strategy*, business *tactics*, business *intelligence*, and so on.

The Second World War

As mentioned, statistics made a big difference during World War II. For instance, the Allied forces accurately estimated the production of German tanks using statistical methods. They also used statistics and logical rules to decode German messages.

The Kerrison Predictor was one of the fully automated anti-aircraft fire control systems that could gun an aircraft based on simple inputs such as the angle to the target and the observed speed. The British Army used this effectively in the early 1940s.

The Manhattan Project was a U.S. government research project in 1942–1945 that produced the first atomic bomb. Under this, the first atomic bomb was exploded in July 1945 at a site in New Mexico. The following month, the other atomic bombs that were produced by the project were dropped on Hiroshima and Nagasaki, Japan. This project used statistics to run simulations and predict the behavior of nuclear chain reactions.

Where Else Was Statistics Involved?

Weather predictions, especially rain, affected the world economy the most since weather affected the agriculture industry. The first attempt was made to forecast the weather numerically in 1922 by Lewis Fry Richardson.

The first successful numerical prediction was performed using the ENIAC digital computer in 1950 by a team of American meteorologists and mathematicians.[2]

Then, 1956 saw analytics solve the shortest-path problem in travel and logistics, radically changing these industries.

In 1956 FICO was founded by engineer Bill Fair and mathematician Earl Isaac on the principle that data used intelligently can improve business decisions. In 1958 FICO built its first credit scoring system for American investments, and in 1981 the FICO credit bureau risk score was introduced.[3]

Historically, by the 1960s, most organizations had designed, developed, and implemented centralized computing systems for inventory control. Material requirements planning (MRP) systems were developed in the 1970s.

In 1973, the Black-Scholes model (or Black–Scholes–Merton model) was perfected. It is a mathematical model of a financial market containing certain derivative investment instruments. This model estimates the price of the option/stock overtime. The key idea behind the model is to hedge the option by buying and selling the asset in just the right way and thereby eliminate risk. It is used by investment banks and hedge funds.

By the 1980s, manufacturing resource planning systems were introduced with the emphasis on optimizing manufacturing processes by synchronizing materials with production requirements. Starting in the late 1980s, software systems known as enterprise resource planning systems became the drivers of data accumulation in business. ERP systems are software systems for business management including models supporting functional areas such as planning, manufacturing, sales, marketing, distribution, accounting, and so on. ERP systems were a leg up over MRP systems. They include modules not only related to manufacturing but also to services and maintenance.

---

[2]http://journals.ametsoc.org/doi/pdf/10.1175/BAMS-89-1-45
[3]www.fico.com/en/about-us#our_history

The Dawn of Business Intelligence

Typically, early business applications and ERP systems had their own databases that supported their functions. This meant that data was in silos because no other system had access to it. Businesses soon realized that the value of data can increase manyfold if all the data is in one system together. This led to the concept of a data warehouse and then an enterprise data warehouse (EDW) as a single system for the repository of all the organization's data. Thus, data could be acquired from a variety of incompatible systems and brought together using extract, transform, load (ETL) processes. Once the data is collected from the many diverse systems, the captured data needs to be converted into information and knowledge in order to be useful. The business intelligence (BI) systems could therefore give much more coherent intelligence to businesses and introduce the concepts of one view of customers and customer lifetime value.

One advantage of an EDW is that business intelligence is now much more exhaustive. Though business intelligence is a good way to use graphs and charts to get a view of business progress, it does not use high-end statistical processes to derive greater value from the data.

The next question that business wanted to answer by the 1990s–2000 was how the data can be used more effectively to understand embedded trends and predict future trends. The business world was waking up to *predictive analytics*.

What are the types of analytics that exist now? The analytics journey generally starts off with the following:

- *Descriptive statistics*: This enables businesses to understand summaries generally about numbers that the management views as part of the business intelligence process.

- *Inferential statistics*: This enables businesses to understand distributions and variations and shapes in which the data occurs.

- *Differences statistics*: This enables businesses to know how the data is changing or if it's the same.

- *Associative statistics*: This enables businesses to know the strength and direction of associations within data.

- *Predictive analytics*: This enables businesses to make predictions related to trends and probabilities.

Fortunately, we live in an era of software, which can help us do the math, which means analysts can focus on the following:

- Understanding the business process

- Understanding the deliverable or business problem that needs to be solved

- Pinpointing the technique in statistics that will be used to reach the solution

- Running the SaaS to implement the technique

- Generating insights or conclusions to help the business

CHAPTER 2

■ ■ ■

Accessing SAS and R

This chapter gives you an introduction to the popular software called SAS and R. It will cover how to install them and get started using them.

Why SAS and R?

Let's first look at the market reality, as mentioned by Gartner in its 2015 report called "Magic Quadrant for Advanced Analytics Platforms." You can find a copy of this report on the Gartner web site at www.gartner.com/technology/research.jsp.

Market Overview

Gartner estimates that the advanced analytics market amounts to more than $1 billion across a wide variety of industries and geographies. Financial services, retail/e-commerce, and communications are probably the largest industries, although use cases exist in almost every industry. North America and Europe are the largest geographical markets, although Asia/Pacific is also growing rapidly.

This market has existed for more than 20 years. The concept of big data not only has increased interest in this market but has significantly disrupted it. The following are key disruptive trends cited by Gartner:

- The growing interest in applying the results of advanced analytics to improve business performance is rapidly expanding the number of potential applications of this technology and its audience across organizations. Rather than being the domain of a few select groups (for example, those responsible for marketing and risk management), every business function now has a legitimate interest in this capability.

- The rapid growth in the amount of available data, particularly new varieties of data (such as unstructured data from customer interactions and streamed machine-generated data), requires greater levels of sophistication from users and systems, as well as the ability to rapidly interpret and respond to data to realize its full potential.

- The growing demand for these types of capabilities is outpacing the supply of expert users, which necessitates higher levels of automation and increases demand for self-service and citizen data scientist tools.

© Subhashini Sharma Tripathi 2016
S. S. Tripathi, *Learn Business Analytics in Six Steps Using SAS and R*, DOI 10.1007/978-1-4842-1001-7_2

What Is Advanced Analytics?

Gartner defines *advanced analytics* as the analysis of all kinds of data using sophisticated quantitative methods (for example, statistics, descriptive and predictive data mining, simulation, and optimization) to produce insights that traditional approaches to business intelligence (BI)—such as query and reporting—are unlikely to discover.

I find this last part to be significant. Advanced analytics is about using methods beyond BI that involve statistics and data mining.

As mentioned, SAS and R are the leaders in the categories of licensed software and free, open source languages, respectively. Thus, if we as analysts can work with both of these languages, we can be assured of being employable for a large set of projects and companies in analytics.

Here are some other points worth noting:

- *The habit of SAS is hard to break*: Traditionally SAS has been the language of analytics, and years of code has been written and perfected in SAS. For an industry to overthrow all of these established processes and start off with R is difficult.

- *Distrust of freeware is high*: Businesses feel comfortable working on products that they pay for and have customer support for. R is free software (though there are many web forums to focus on it). Tech support is available for paid versions such as Revolution Analytics. (Refer to www.revolutionanalytics.com/why-revolution-analytics for more information on the consulting and tech support services from Revolution Analytics.)

- *R has in-memory processing*: Since R works on in-memory processing, there are several issues related to big data processing. However, enterprise versions and RHadoop have offset these limitations. (Enterprise versions of R are not free.)

- *Coding intensity is higher in R while SAS has invested in a lot of point-and-click interfaces such as E Miner and EG*. SAS also has many customized suits for specific business requirements and functions, making it easier to deploy.

Industry-specific solutions exist for the following industries in SAS:

- Automotive
- Banking
- Capital markets
- Casinos
- Communications
- Consumer goods
- Defense and security
- Government
- Healthcare providers
- Health insurance
- High-tech manufacturing
- Higher education
- Hotels

- Insurance

- K-12 education

- Life sciences

- Manufacturing

- Media

- Oil and gas

- Retail

- Small and midsize business

- Sports

- Travel and transportation

- Utilities

■ **Tip**   Read more at the SAS web site at `www.sas.com/en_us/industry.html`.

History of SAS and R

I am sure you are curious to understand how SAS and R evolved. Let's look at their histories.

History of SAS

SAS is definitely the tried-and-tested superstar of the analytics industry. In 1966 there was a need for a computerized statistics program to analyze agricultural data collected by the U.S. Department of Agriculture. The U.S. Department of Agriculture was funding the research for a consortium of eight land-grant universities, and these schools came together under a grant from the National Institute of Health to develop a general-purpose statistical software package for the analysis of agricultural data to improve crop yield. The resulting program was called the *statistical analysis system*, and the acronym SAS arose from the name.

Out of the eight universities, North Carolina State University became the leader of the consortium because it had access to a more powerful mainframe computer compared to other universities.

North Carolina State University faculty members Jim Goodnight and Jim Barr were the project leaders. When the National Institute of Health discontinued funding in 1972, members of the consortium agreed to chip in money each year to allow North Carolina State University to continue developing and maintaining the system and supporting the statistical analysis needs. In 1976, the team working on SAS took the project out of the university and incorporated the SAS Institute. In 1985, SAS was rewritten in the C programming language, and the science enterprise Miner was released in 1999. As the name suggests, it was the start of SAS creating suites of products for solving specific business problems, whereas Enterprise Miner was aimed at mining large data sets. In 2002, the Text Minor software was introduced. Today SAS products include the following:

- SAS 9.4 (base SAS)

- SAS/STAT

- SAS Analytics Pro

- SAS Curriculum Pathways

- SAS Data Management

- SAS Enterprise Miner

- SAS Marketing Optimization

- SAS University Edition

- SAS Visual Analytics

- SAS Visual Statistics

What I will cover here is base SAS (which will enable you to write code in SAS), and I will use SAS Enterprise Guide (EG) as the platform so that you also get exposure to the point-and-click functionalities.

What Is EG?

SAS Enterprise Guide provides an intuitive project-based programming and point-and-click interface to SAS. It includes an intelligent program editor, querying capabilities, repeatable process flows, stored process creation and consumption, and a multitude of other features. It allows for point-and-click tasks and the editing of the code for these tasks. Thus, it allows for much less code writing. As an analyst, if you understand the construct of the code and can edit the code to create customized outputs, the time savings is huge. Also, you break up the repetition and monotony.

The other benefit is that noncoders can work more efficiently. Thus, people like me are much more comfortable using it.

How Can You Access SAS Enterprise Guide Software?

SAS has created a SAS on-demand facility for academic users and students. View it and install SAS on your system by visiting http://support.sas.com/software/products/ondemand-academics/#s1=2.

Otherwise, just Google *SAS on Demand for Academics* and you will be able to see all the relevant links.

■ **Tip**   SAS continues to update versions of the software and sometimes the look and feel of the SAS on-demand site. Don't be surprised to see changes. It's a good way to get an idea of new products or new versions of products that SAS releases.

History of R

In 1975–76, Bell Laboratories designed S, which is a statistical computing language, as an alternative to the common statistical computing that was done by directly calling Fortran subroutines.

In 1995, Ross Ihaka and Robert Gentleman wrote an experimental R, which was "not unlike S." In the last two decades, R has emerged as a software application for statistics, data management, programming, and so on, which exists in a unique quantity and variety. The quality varies, but on average the output is impressive. Most of this is in an open environment that encourages improvements and has wide participation from the statistics profession.

R is freely available under the GNU General Public License; check out www.r-project.org/.

Why Was R Named R?

R was so named partly because of the initials of the founders, Ross Ihaka and Robert Gentleman, and partly because it was a play on the name of the language S. Interesting, right? You can look up the 2015 R FAQ at http://cran.r-project.org/doc/FAQ/R-FAQ.html for more details.

What Is R?

R is a system for statistical computation and graphics. It consists of a language plus a runtime environment with graphics, a debugger, access to certain system functions, and the ability to run programs stored in script files.

The design of R has been heavily influenced by two existing languages: Becker, Chambers, and Wilks' S (see "What Is S?" in the previously mentioned FAQ) and Sussman's Scheme. Whereas the resulting language is similar in appearance to S, the underlying implementation and semantics are derived from Scheme. See "What Are the Differences Between R and S?" in the previously mentioned FAQ for further details.

The core of R is an interpreted computer language that allows branching and looping as well as modular programming using functions. Most of the user-visible functions in R are written in R. It is possible for the user to interface to procedures written in C, C++, or Fortran for efficiency. The R distribution contains functionality for a large number of statistical procedures. Among these are linear and generalized linear models, nonlinear regression models, time-series analysis, classical parametric and nonparametric tests, clustering, and smoothing. There are also a large set of functions, which provide a flexible graphical environment for creating various kinds of data presentations. Additional modules (called *add-on packages*) are available for a variety of specific purposes (see "Add-on Packages in R" in the previously mentioned FAQ).

What Is RStudio?

The R console is a great fit for programmers and coders, but for noncoders (people like me), it's easier to download and use RStudio. RStudio is a front end to R, as shown in Figure 2-1. You need R to make RStudio work. RStudio makes using R a lot easier and smoother and lets you use lots of packages easily.

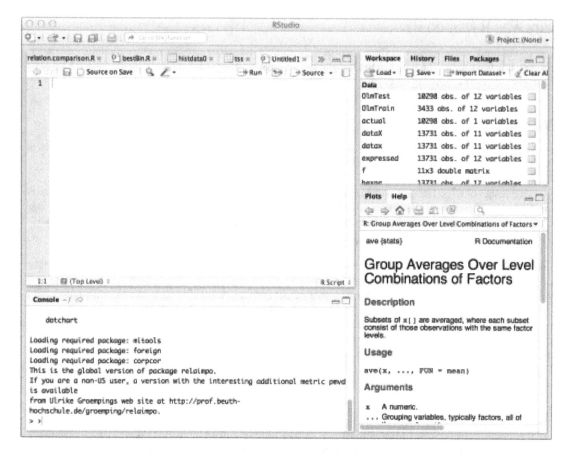

Figure 2-1. *RStudio*

The Help pane displays the answers to your queries. The upper-right Workspace pane lists the data, values, and functions in the current workspace. The Import Dataset button helps you write the read command for parsing input data files (CSVs or files with other delimiters). It gives a handy preview of the resulting R object.

RStudio can be used over the Internet. To launch RStudio, go to `http://beta.rstudio.org` and log in using your Gmail address.

Read more about it and download it from `www.rstudio.com/`.

■ **Tip** Download R and then RStudio for a smooth installation on your computer. For server-based R, download the RStudio Server version (`www.rstudio.com/products/rstudio/download-server/`).

What Is CRAN?

The Comprehensive R Archive Network (CRAN) is a collection of sites that carry identical material, consisting of the R distributions, the contributed extensions, the documentation for R, and the binaries.

The CRAN master site at Wirtschaftsuniversität Wien (WU) in Austria is at this location:

```
http://CRAN.R-project.org/
```

Daily mirrors are available at these URLs:

http://cran.at.R-project.org/ (Wirtschaftsuniversität Wien, Austria)
http://cran.au.R-project.org/ (University of Melbourne, Australia)
http://cran.br.R-project.org/ (Universidade Federal do Paraná, Brazil)
http://cran.ch.R-project.org/ (ETH Zürich, Switzerland)
http://cran.dk.R-project.org/ (dotsrc.org, Aalborg, Denmark)
http://cran.es.R-project.org/ (Spanish National Research Network, Madrid, Spain)
http://cran.pt.R-project.org/ (Universidade do Porto, Portugal)
http://cran.uk.R-project.org/ (U of Bristol, United Kingdom)

See http://CRAN.R-project.org/mirrors.html for a complete list of mirrors. Please use the CRAN site closest to you to reduce network load.

From CRAN, you can obtain the latest official release of R, daily snapshots of R (copies of the current source trees) as gzipped and bzipped tar files, a wealth of additional contributed code, and prebuilt binaries for various operating systems (Linux, Mac OS Classic, OS X, and Microsoft Windows). CRAN also provides access to documentation on R, existing mailing lists, and the R bug tracking system.

Which Add-on Packages Exist for R?

Packages are collections of R functions, data, and compiled code in a well-defined format. They help users to use code related to certain statistical and visualization techniques. The directory where packages are stored is called the *library*. R has some standard packages and some that have to be downloaded and installed. Once installed, they have to be loaded into the session to be used, either by code commands or with the GUI in RStudio.

The R distribution comes with the following packages:

- base: Base R functions (and data sets before R2.0.0)

- compiler: R bytecode compiler (added in R2.13.0)

- datasets: Base R data sets (added in R2.0.0)

- grDevices: Graphics devices for base and grid graphics (added in R2.0.0)

- graphics: R functions for base graphics

- grid: A rewrite of the graphics layout capabilities, plus some support for interaction

- methods: Formally defined methods and classes for R objects, plus other programming tools, as described in the Green Book

- parallel: Support for parallel computation, including by forking and by sockets, and random-number generation (added in R2.14.0)

- splines: Regression spline functions and classes

- stats: R statistical functions

- stats4: Statistical functions using S4 classes

- tcltk: Interface and language bindings to Tcl/Tk GUI elements

- tools: Tools for package development and administration

- utils: R utility functions

■ **Tip** To view details of add-on packages, I recommend using Google to find what you want and then check out the comments at http://stackoverflow.com.

R is highly conducive to visualization functions and has dedicated packages such as ggplot 2, which produces high-quality plots that help develop ideas to build models or do advanced analytics. Also, the turnaround time of new methodologies or algorithms being discovered and implemented in R is much shorter than that for licensed software like SAS.

One large drawback is that the syntax is inconsistent across packages, making the learning process more difficult.

▦ **Note** Microsoft has acquired Revolution Analytics. Read more at http://blogs.technet.com/b/machinelearning/archive/2015/04/06/microsoft-closes-acquisition-of-revolution-analytics.aspx.

Why Is Microsoft's Acquisition of Revolution Analytics Important?

Revolution Analytics is the leading commercial provider of software and services based on R, and Microsoft is planning to build R into SQL Server, according to Joseph Sirosh, Microsoft's corporate VP of information management and machine learning. This will enable customers to deploy it in a data center, on Azure, or in a hybrid configuration. Microsoft also plans to integrate Revolution's R distribution into Azure HDInsight and Azure Machine Learning to make it easier to analyze big data.

Ideally this will make the R interface much simpler and user-friendly and the infrastructure as a service (IaaS) Microsoft Azure much easier to use too.

Installing SAS and R

Let's install the software.

Installing SAS

As mentioned earlier, you can use the resources provided by SAS for teaching and training. There are three ways to access SAS (www.sas.com/en_us/learn/analytics-u.html).

- SAS University Edition
- SAS OnDemand for Academics
- Education Analytical Suite

SAS University Edition

This is best for teaching and learning SAS skills and analyzing data using SAS foundational technologies. You have these choices:

- You can download it from SAS.
  - It's free!
  - It runs on Windows, Linux, and Mac.
  - It runs locally. No Internet connection is needed.

- You can access it via the Amazon Web Services (AWS) marketplace.

    - It's free SAS software (AWS usage fees may apply).
    - It runs in the cloud; all you need are a browser and an Internet connection.

Let's start with downloading from SAS.

1. Go to www.sas.com/en_us/software/university-edition.html.

2. Click "Get free software" (Figure 2-2).

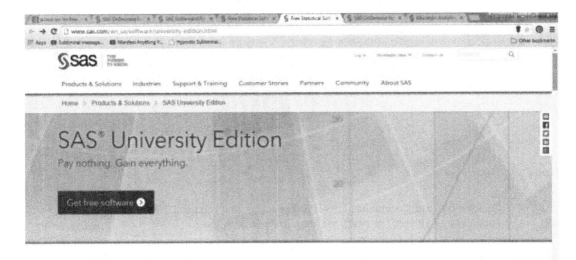

Figure 2-2. Click to get the free software

3. Click "Download now" (Figure 2-3).

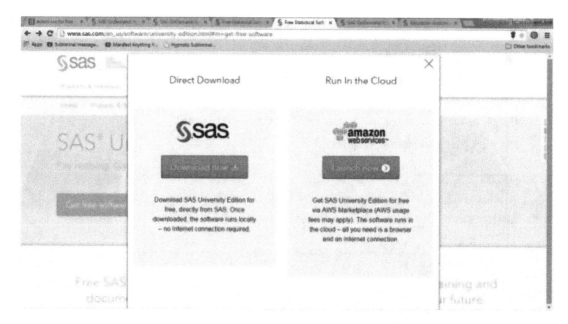

Figure 2-3. *The two options*

4. Click the compatible virtualization software package that you need (for example, I need VMware Player 7 or later), as shown in Figure 2-4.

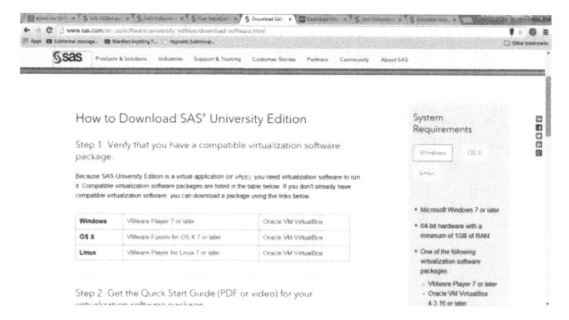

Figure 2-4. *The virtualization choices*

5. Click the relevant package that you need (for example, I need VMware Player for a Windows 64-bit operating system). How will you know if your operating system is 32-bit or 64-bit? Go to the Control Panel of the system and open System and Security; you will find it under System. For example, on my machine, the information looks like Figure 2-5.

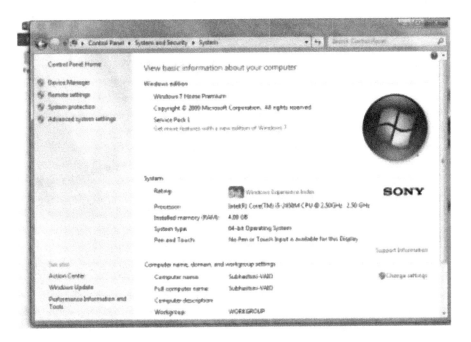

Figure 2-5. *The Control Panel*

6. Run the VMware EXE (which will be visible in your Downloads folder or any folder in which you download it) and complete the installation.

7. Go to the SAS University Edition's Quick Start Guide for VMware Player (Figure 2-6). For me the guide is at http://support. sas.com/software/products/university-edition/docs/en/ SASUniversityEditionQuickStartVMwarePlayer.pdf.

Figure 2-6. *Quick Start Guide location*

8. Find where the SAS software is located (Figure 2-7).

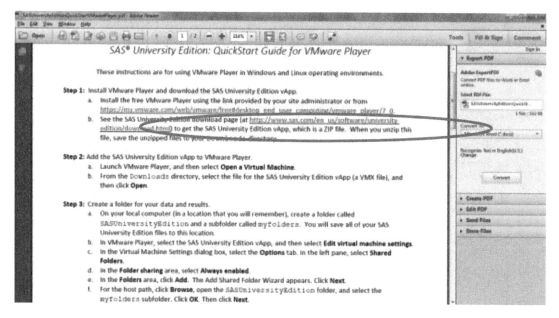

Figure 2-7. *The Quick Start Guide shows where the software is*

9. Download the software.

10. Log in with your SAS profile (Figure 2-8).

Figure 2-8. *Logging into SAS*

11. Confirm the terms and conditions.

12. Start the download from the receipt page (Figure 2-9).

Figure 2-9. Downloading the SAS University Edition

■ **Tip**   If you see the error "This kernel requires an x86-64 CPU, but only detected an i686 CPU. Unable to boot – please use a kernel appropriate for your CPU," then enable Intel VT-x/AMD-V from BIOS. Google this to find some online help on this issue.

SAS OnDemand for Academics

Using the SAS OnDemand for Academics site is best for gaining online access to the powerful SAS software via the cloud, typically as part of an academic course. Here are the benefits:

- It's free.

- It runs on Windows, Linux, and Mac.

- It's accessible via the cloud whenever and wherever there's an Internet connection.

- The available data storage is up to 5GB.

Follow the step-by-step installation and access guide at www.sas.com/en_us/industry/higher-education/on-demand-for-academics.html.

1.   Choose the option Independent Learners (Figure 2-10).

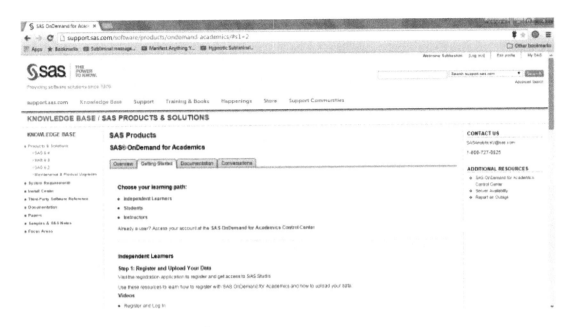

Figure 2-10. *Choose the learning path*

2. Download the Quick Start Guide, as shown in Figure 2-11.

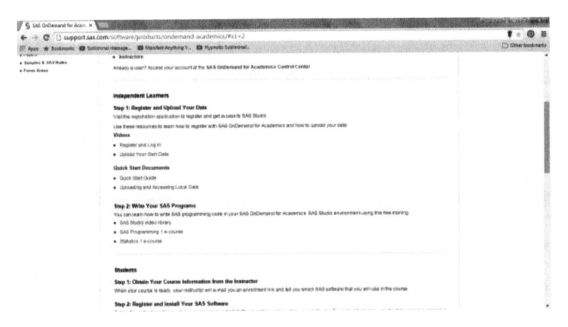

Figure 2-11. *Get the Quick Start Guide*

3. Register your account (Figure 2-12).

Figure 2-12. *Registering*

4. Log in to SAS Studio and begin (Figure 2-13).

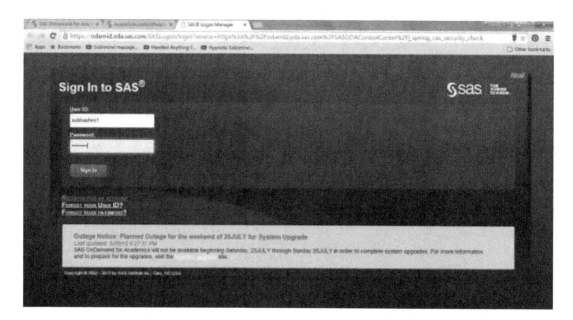

Figure 2-13. *Logging into SAS Studio*

5. Choose Enterprise Guide and follow the installation process (Figure 2-14).

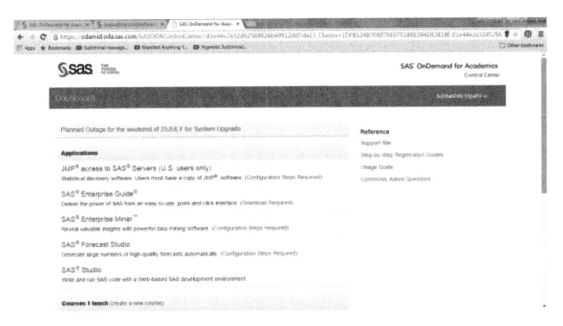

Figure 2-14. *The Dashboard*

■ **Tip** Store your login ID (which the SAS system generates at registration) carefully. If you forget it, you may have to face unnecessary hassle when logging in next time.

Education Analytical Suite

This way is best for institutions wanting in-house software and data for teaching and academic research. It provides the comprehensive SAS foundational technologies via a reduced-cost enterprise license. Here are the benefits:

- Flexible, low-cost, and unlimited licenses

- Runs on Windows, Linux, and others

- Runs locally (no Internet connection needed)

- Local, unlimited data storage

Follow the information given at `www.sas.com/en_us/industry/higher-education/education-analytical-suite.html`.

Installing R

You need to install RConsole and RStudio for the complete experience, as discussed earlier. You will notice that installing R is much simpler after you have SAS installed.

1. Download the relevant version from www.r-project.org/ (Figure 2-15).

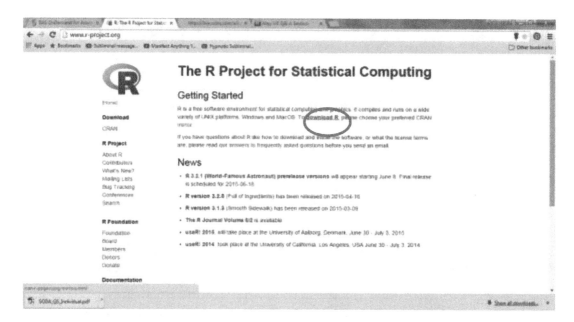

Figure 2-15. *Choosing to download R*

2. Choose the CRAN mirror closest to you at http://cran.r-project.org/ mirrors.html (Figure 2-16).

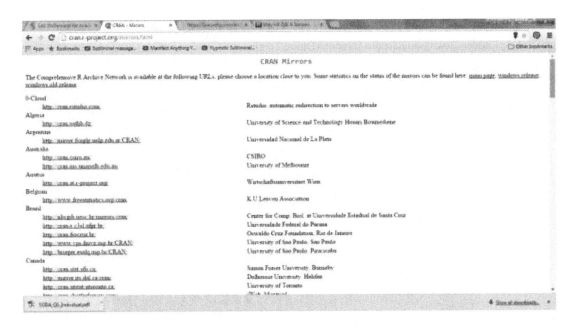

Figure 2-16. *A list of available mirrors*

3. Download and install R for the type of computer you use (Figure 2-17).

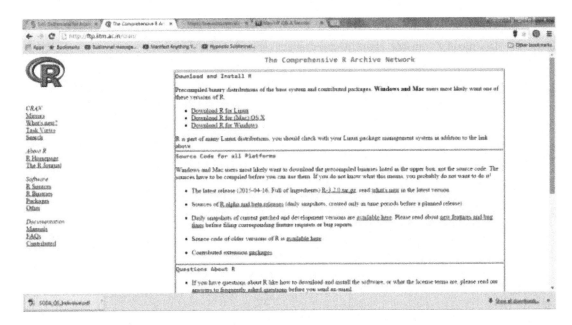

Figure 2-17. *R versions*

4.   Download RStudio from www.rstudio.com/products/rstudio/.

5.   Choose Download RStudio Desktop (Figure 2-18).

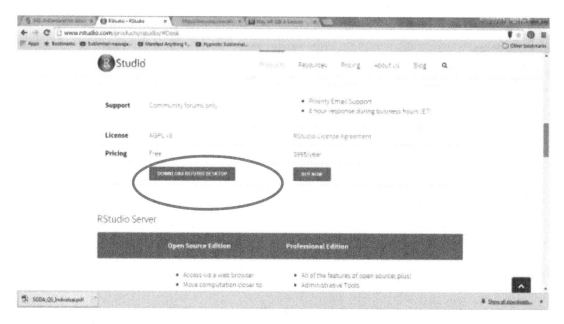

Figure 2-18. *Choosing RStudio Desktop*

6.   Download it from www.rstudio.com/products/rstudio/download/ (Figure 2-19).

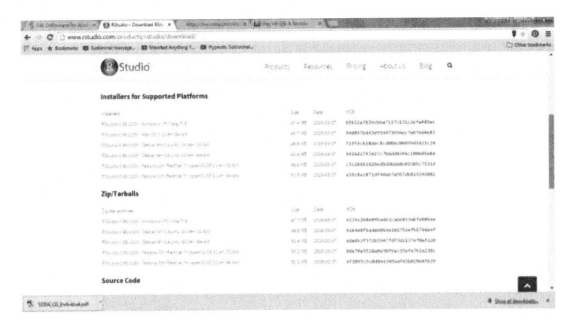

Figure 2-19. *All the installers*

Now you are good to go. Let's start doing analytics with SAS and R!

CHAPTER 3

■ ■ ■

Data Manipulation Using SAS and R

In this chapter, I will cover the following:

- The data flow from ERP to business analytics SaaS
- How to do a sanity check on data
- How to merge data sets
- How to deal with missing values, duplication, and outliers
- How to create a project datamart

Define: The Phase Before Data Manipulation (Collect and Organize)

Let's talk a bit about the "define" in DCOVA&I. As you start a project, you need some clear directions on what you intend to "solve for." After all, the project will be executed. The clear direction in terms of an analytics project is what you seek to get into the define phase. Since you need to use statistical techniques later, the analyst needs to understand the business problem and then convert it to a statistical problem. This may not turn out to be as simple as it seems. Invariably, the business comes up with a business word or terminology-intensive problem statement. For example, you may get a request from the CEO's office saying something like this: "Please look at the data and tell me who are this company's most profitable customers." To convert this business problem statement into an analytical problem statement, the analyst will employ the following process (see Figure 3-1):

1. Ask for the definition in business terms from the business. Who, as per the business, is a profitable customer ? Someone who generates revenues of the hundreds? Or someone in at least the thousands ?

2. Create your own understanding of the problem by using your knowledge of the following:

    - Industry (banking, retail, manufacturing, and so on)
    - Process (marketing, risk, operations , customer relationship management [CRM], and so on)
    - Data types commonly associated with the industry and process

© Subhashini Sharma Tripathi 2016
S. S. Tripathi, *Learn Business Analytics in Six Steps Using SAS and R*, DOI 10.1007/978-1-4842-1001-7_3

3. Convert your understanding of the problem to a few possible mathematical problems (business $y's$).

4. Propose your solutions to the business.

5. Help the business modify/choose the one that will be used for the project.

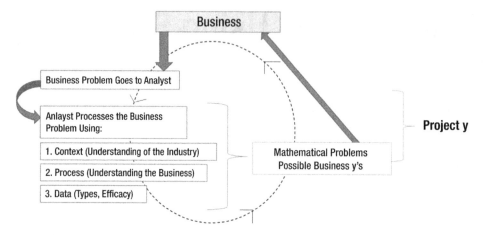

Figure 3-1. *The define process: identifying the y for the project*

Basic Understanding of Common Business Problems

What is a business? It's an organization or economic system where goods and services are exchanged for money. Businesses require some form of investment and enough customers to whom its output can be sold on a consistent basis in order to make a profit. The goal, therefore, of a business is to have higher sales than expenditures, resulting in a profit.

The efficient and effective operation of a business is called *management*. Owners may administer their businesses themselves or employ managers to do this for them.

Management in business organizations is the function that coordinates the efforts of people to accomplish goals and objectives using the available resources efficiently. Management includes planning, organizing, staffing, leading, controlling, and organizing for the company to accomplish its goals. For most companies, the primary goal is a profit (Figure 3-2).

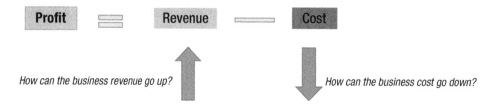

Figure 3-2. *The goal of commercial business is maximum profits*

Strategic objectives are one of the fundamental building blocks of a company's strategic plan. Financial strategic objectives look like this:

*   To exceed $10 million in sales in the next ten years.

*   To increase revenue by 10 percent annually.

Customer-centric strategic objectives look like this:

- Increase customer retention to 20 percent.

- Expand sales to four countries in the next two years.

Operational strategic objectives look like this:

- Capitalize on physical facilities such as location, and so on.

- Develop and implement a promotional plan to drive increased business.

People-related strategic objectives look like this:

- Employ professionals who create success for customers.

- Align incentives and staff reports with performance.

Thus, companies invariably create strategic objectives that give a general direction in which the company wants to move to improve its profitability. I have often found that the business problem that the analytics team gets to solve is linked to the strategic objectives of the organization.

It is claimed that 50 percent of analytics projects fail to deliver.

What is failure for an analytics project? It could mean that the analysis was inconclusive, the model was inefficient, the work did not finish in time, the project was not implemented in the business, and so on. One of the biggest reasons for this is often found in the define stage. Thus, for the overall success of the project and for the business to be able to drive the strategic objectives through analytics, it is imperative that stage D—define—should be successfully completed and a sign-off done by all involved.

■ **Tip**   As much as 20 percent of the project time can be spent in the define stage. It is not advisable to rush this stage in the interest of the success of the project.

Sources of Data

As you know, the data available to an organization for analytics can be divided into these two broad categories:

- Secondary data

- Primary data

Secondary data is data that someone else has collected. Common sources of secondary data are government institutions, research organizations, regulatory authorities, consulting firms, and so on.

Some of the common problems with secondary data is that it may not have been collected for the specific purpose of the analysis that you intend to do. The time frame may also not match. Can you really make organization policies on the basis of data that is not specific to the customers/business segments of the organization?

If you look at data sets, you may not be able to answer some questions, such as could there be any missing information in some observations? If yes, will bias be built because of this incompleteness or an ignorance of the incompleteness?

Of course, the advantages are that someone has already found data and you can take advantage of it. Secondary data is often cheaper in terms of the money that is spent on collecting it. It often saves a lot of time and definitely shortens the time difference between the accumulation of data and the analysis. Thus, it may have great exploratory value and can help formulate definitions/questions for your projects and also, perhaps, formulate some hypotheses to test.

Primary data, on the other hand, is the data are that you/the organization collects first-hand. Sources of primary data could be focus groups, questionnaires, or personal interviews, but, in large organizations it is primarily the internal ERP systems. Thus, primary data is data about your customers, vendors, and internal processes.

It has the characteristic of uniqueness. But this may cause a problem because it would not be easy to compare primary data to other populations from outside the organization. However, it is also true that most organizations in an industry would be tracking similar types of primary data. Thus, companies that work in the similar space would track similar variables in their data.

The Use of Benchmarks to Create an Optimal Define Statement

How do you figure out what exactly to aim at when you start a project?

Perhaps for all your projects you will end up using internal data or primary data (Figure 3-3). However, you can use secondary data to set the benchmarks that can help you define what is a good/bad value for your business. For example, if the turnaround time for orders is 30 hours, what is a good per-month revenue figure for this business? This is a subjective question. Everyone in the business will have an opinion on this. How can you really arrive at what will work or what should work for your business? Therefore, an effective use of secondary data is benchmarking. It allows you to look at the other players and industry trends and set standards/deliverables for projects.

■ **Tip**   Look at journals and industry-specific articles from authorities who are expected to be knowledgeable in the industry to arrive at a good benchmark.

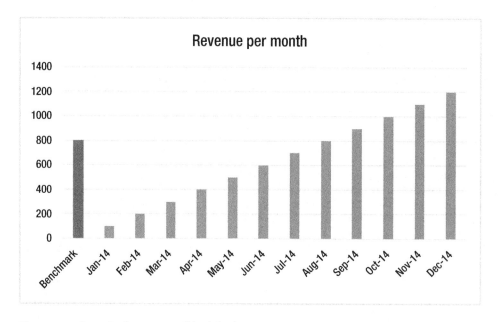

Figure 3-3. *Data for the project will be (often) primary data, while the benchmark will be (often) secondary data*

Data Flow from ERP to Business Analytics SaaS

Let's take a look at where you will receive the data on which you will do the analytics (Figure 3-4).

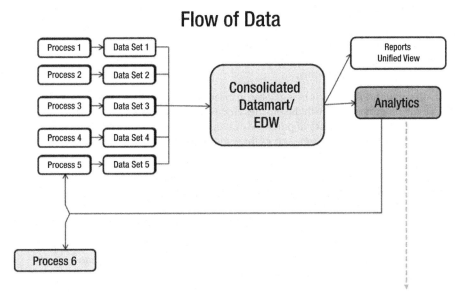

Figure 3-4. Flow of primary data in organizations

An enterprisewide data warehouse (EDW) is the repository of data from different systems present within the organization. The way to link the data across multiple systems is with the help of primary keys. Thus, the EDW can contain data that may link multiple transactions and activities together for the customer, vendor, product, and so on.

What Are Primary Keys?

A primary key is a unique identifier for an entity in a table or database. It helps you find an entity (customer, vendor, or product) easily on your system. A primary key is a variable in a relational database that is unique for each record, customer, vendor, or transaction. Generally, for customers, the customer ID that is generated by the organization is a primary key, or the phone number/mobile number could be a primary key. E-mail IDs, PAN card numbers, Adhaar card numbers, Social Security numbers, employee ID numbers, and so on, are other common values that can be used as primary keys.

What Is a Relational Database?

A relational database or table is so organized as to recognize relationships between stored information. Say an employee wants to see the amount accumulated in his Provisent Fund (PF) account. How will the HR manager check? The HR manager will ask for the employee ID (primary key) and use that to check in the database. Thus, a relational database works via primary keys.

Sanity Check on Data

A development data set is the set of data of a particular timeframe, product, or business that is used for the purpose of analytics and building the model.

The development data set, therefore, could belong to a specific time period. This is because of the following:

- The business as it exists today could be consistent only in the last two years to five years, and so on, since every business undergoes changes and sometimes these changes alter the business significantly.

- It is also possible that the analytics is being done on a variable that can be derived only once the customer has been with the business for a certain period of time. Thus, if you want to measure the frequency of purchase of a customer in an online portal, you will have to look at the data in some sets of three months, six months, one year, and so on.

The data that belongs to this "relevant" time frame is called the *in-time sample data*.

What about the data that is not in this time frame? That data set is called the *out-of-time sample data*. The model that is built with in-sample data can be validated on the out-of-sample data.

What is the sanity check that has to be done on the data? Usually, it may involve any or all of the following:

- Checking whether the appropriate numbers of rows and columns/observations and variables have been imported into the analytics system. You can check this against the source system.

- Checking the formats of the various variables in the data set.

- Checking for missing data.

Let's see how to do this in SAS and R.

■ **Hint** To pick up basic coding, try the following web sites. For SAS, visit `https://support.sas.com/documentation/onlinedoc/guide/tut42/en/menu.htm`. For R, visit `http://tryr.codeschool.com/levels/1/challenges/1`.

Case Study 1

An IT-enabled services (ITES) company wants to understand its data related to service requests for customers. These requests are related to particular products that the company manufactures. The requests can be divided into three priorities: Low, Medium, and High. Each of these priority levels has a different service level agreement (SLA) for resolution.

The data consists of the following fields:

- *ServiceRequestNo*: This is the unique ID for each service request logged on the system.

- *ServiceRequestStatus*: As of a certain date, this is the status of the service request.

- *TypeOfEnagagement*: This is the type of work the company is doing for the client.

- *Incident/Problem*: This specifies whether the service request is a problem.

- *SR Priority*: This is the priority level of the service request.

- *SR Open Date*: This is the date on which the service request was logged in the system.

- *SR Close Date*: This is the date on which the service request was closed in the system.

- *Product* : This is the name of the product for which the service request was raised.

- *Geography*: This is the continent in which the client belongs.

- *Country*: This is the country to which the client belongs.

Let's create a project plan along the lines of DCOVA and I.

1. First you define the problem. Clean the data and create a project datamart to do analytics for Resolution Time, in other words, the time, in a number of days, taken to resolve the service request (to be calculated as the difference between SR Close Date and SR Open Date).

   - Create the *y* variable.

2. Collect the relevant data. The data for the project comes from one file, CaseStudy1.csv.

3. Organize the data. Manipulate data, create derived variables through calculation, and understand the missing values.

4. Visualize the data.

   - Univariate analysis of *y*

   - Multivariate analysis: correlation

5. Create the final project datamart.

   - Drop the variables used to create *y* since these variables have been used to create the business problem. Therefore, they don't need to be reused. If you use them again, it will be double counting.

   - Drop the variables that are non-numeric. Statistics and math can be done only on numeric variables.

   - Drop these numeric values: ServiceRequestStatus, TypeOfEnagagement, Incident/Problem, SR Priority, Product, Geography, and Country.

Case Study 1 with SAS

Now you'll execute the previously mentioned steps in SAS.

1. Create libname. In SAS OnDemand for Academics, the library is preset for the course. You can view the name via the SAS Studio link. (Hint: Google is your best friend when you want to find relevant documents and do-it-yourself processes. SAS has extensive online help.)

```
libname lib1 " /home/subhashini1/my_content "; run;
PROC CONTENTS data = lib1._ALL_ NODS;run;
```

Page Break

The CONTENTS Procedure

Directory	
Libref	LIB1
Engine	V9
Physical Name	/courses/d4660de5ba27fe300
Filename	/courses/d4660de5ba27fe300
Inode Number	10354689
Access Permission	rwxr-xr-x
Owner Name	subhashini1
File Size	4KB
File Size (bytes)	4096

Page Break

2. Import the data. I loaded the CSV file into the SAS on-demand cloud directory.

```
FILENAME REFFILE "/home/subhashini1/my_content/CaseStudy1.csv" TERMSTR=CR;

PROC IMPORT DATAFILE=REFFILE
        DBMS=CSV
        OUT=WORK.IMPORT;
        GETNAMES=YES;
RUN;
```

■ **Note** An easier way to import without touching the cloud-based server is to use the button-driven option File ➤ Import Data. This will open an Import Wizard through which you can pull the data into the SAS system from your desktop.

Did the file get loaded? Is all the data present? PROC CONTENTS will help you determine this.

PROC CONTENTS DATA=WORK.IMPORT; RUN;

The CONTENTS Procedure

Data Set Name	WORK.IMPORT	Observations	2016
Member Type	DATA	Variables	10
Engine	V9	Indexes	0
Created	08/10/2015 14:48:15	Observation Length	80
Last Modified	08/10/2015 14:48:15	Deleted Observations	0
Protection		Compressed	NO
Data Set Type		Sorted	NO
Label			
Data Representation	SOLARIS_X86_64, LINUX_X86_64, ALPHA_TRU64, LINUX_IA64		
Encoding	utf-8 Unicode (UTF-8)		

Engine/Host Dependent Information

Data Set

Alphabetic List of Variables and Attributes

#	Variable	Type	Len	Format	Informat
10	Country	Char	9	$9.	$9.
9	Geography	Char	12	$12.	$12.
4	Incident/ Problem	Char	8	$8.	$8.
8	Product	Char	4	$4.	$4.
7	SR Close Date	Num	8	DATE8.	DATE8.
6	SR Open Date	Num	8	DATE8.	DATE8.
5	SR Priority	Char	8	$8.	$8.
1	ServiceRequestNo	Char	5	$5.	$5.
2	ServiceRequestStatus	Char	6	$6.	$6.
3	TypeOfEnagagement	Char	8	$8.	$8.

■ **Note** Check out the *Little SAS Book for Enterprise Guide* (purchase it from the SAS web site at https://www.sas.com/store/prodBK_61861_en.html). For free help on SAS coding, visit www.sascommunity.org/wiki/Sasopedia/Topics.

3. Store the SAS data set back in the SAS on-demand library.

    DATA LIB1.CS1;
    SET WORK.IMPORT; RUN;

Program* | Log | Output Data | Results
Export ▾ Send To ▾ Create ▾ | ⬆ ⬇ | Log Summary | Project Log | Properties

```
DATA LIB1.CS1;
SET WORK.IMPORT; RUN;


There were 2016 observations read from the data set WORK.IMPORT.
The data set LIB1.CS1 has 2016 observations and 10 variables.
DATA statement used (Total process time):
real time              0.07 seconds
```

4. View five lines from the SAS file CS1 (for Case Study 1) to see how the data looks.

    PROC PRINT DATA=LIB1.CS1(OBS=5);RUN;

#	Name	Member Type	File Size	Last Modified
1	CS1	DATA	384KB	08/10/2015 09:37:22

Page Break

Obs	ServiceRequestNo	ServiceRequestStatus	TypeOfEnagagement	Incident/ Problem	SR Priority	SR Open Date	SR Close Date	Proc
1	1001	Closed	Project	Incident	2-High	02JAN14	01MAR14	APC
2	1002	Closed	Project	Incident	2-High	02JAN14	01MAR14	APC
3	1003	Closed	Contract	Problem	2-High	23JAN14	22OCT14	APC
4	1004	Closed	Contract	Incident	3-Medium	24JAN14	24MAR14	APC
5	1005	Closed	Contract	Problem	3-Medium	28JAN14	23APR14	APC

5. Work on the file WORK.IMPORT in the WORK directory. Then save the final, modified data set in the permanent directory (LIB1). Create a copy of the file WORK.IMPORT as WORK.CS1.

    DATA WORK.CS1;
    SET WORK.IMPORT; RUN;

```
DATA WORK.CS1;
SET WORK.IMPORT; RUN;
```

```
There were 2016 observations read from the data set WORK.IMPORT.
The data set WORK.CS1 has 2016 observations and 10 variables.
DATA statement used (Total process time):
real time              0.00 seconds
```

6. Create the *y* variable of Resolution Time, which is the difference between SR Close Date and SR Open Date.

```
VALIDNAME=ANY;
DATA WORK.CS1;
SET WORK.CS1;
RESOLUTIONTIME= 'SR Close Date'n-'SR Open Date'n; RUN;

PROC PRINT DATA=WORK.CS1(OBS=5);RUN;
```

FAQ   Why do you use VALIDNAME=ANY; and n with the variable names 'SR Close Date'n-'SR Open Date'n? SAS has strict naming conventions. A valid name for SAS has to follow these conventions:

Names can be up to 32 characters in length.

The first character must begin with an English letter or an underscore. Subsequent characters can be English letters, numeric digits, or underscores.

A variable name cannot contain blanks.

A variable name cannot contain any special characters other than the underscore.

A variable name can contain mixed case. When SAS processes variable names, however, it internally uppercases them. For example, city, City, and CITY all represent the same variable.

To override these conventions, you write the command VALIDNAME=ANY.

There are spaces in the variable names that will not be acceptable to SAS. Ideally, during import, any spaces or special characters in variable names should be replaced with an underscore (_) in SAS.

FAQ   How do you know the variable name in the SAS data set? You can see the names and type of variable, whether character, date, or numeric, in the PROC CONTENTS output.

---

7. Let's understand the distribution of the *y* variable.

    Choose the Task tab and then the Describe data tab and then choose Distribution Analysis.

    Choose Normal Distribution and Histogram Plot.

■ **Tip**   Click Preview Code to see the SAS code for the options you have chosen.

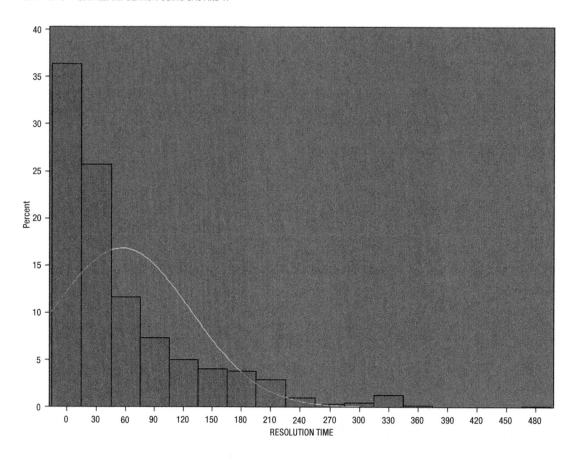

8. You will see the univariate analysis of the *y* variable. What do the descriptive stats say?

■ **Note** Use the `Proc Unvariate` procedure to run similar output in code.

The UNIVARIATE Procedure
Variable: RESOLUTIONTIME

Moments

N	1791	Sum Weights	1791
Mean	56.4204355	Sum Observations	101049
Std Deviation	71.083311	Variance	5052.8371
Skewness	1.87603057	Kurtosis	3.84706109
Uncorrected SS	14745807	Corrected SS	9044578.41
Coeff Variation	125.98859	Std Error Mean	1.67965411

Basic Statistical Measures

Location		Variability	
Mean	56.42044	Std Deviation	71.08331
Median	28	Variance	5053
Mode	0	Range	484
		Interquartile Range	74

Tests for Location: Mu0=0

Test	Statistic		p Value	
Student's t	t	33.59051	Pr > |t|	<.0001
Sign	M	809.5	Pr >= |M|	<.0001
Signed Rank	S	655695	Pr >= |S|	<.0001

Quantiles (Definition 5)

Level	Quantile
100% Max	484
99%	320
95%	206
90%	161
75% Q3	80
50% Median	28
25% Q1	6
10%	1
5%	0
1%	0
0% Min	0

Extreme Observations

Lowest		Highest	
Value	Obs	Value	Obs
0	2006	347	560
0	1991	348	1386
0	1969	349	518
0	1929	478	564
0	1920	484	1838

Missing Values

Missing		Percent Of	
Value	Count	All Obs	Missing Obs
.	225	11.16	100

■ **Tip** Copy and paste the outputs from the Results tab into an Excel sheet. You will find that the format fits perfectly in Excel.

■ **Note** Code in SAS is not case-sensitive. You can write code in uppercase or lowercase and it will run with equal effectiveness.

9. Drop the observations with *y* missing.

```
DATA WORK.CS2;
SET WORK.CS1;
IF RESOLUTIONTIME >=0; RUN;
```

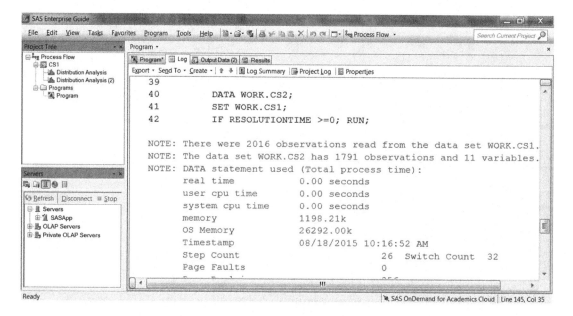

10. Do some visualizations to understand the data.

   Run Tables from Tasks ➤ Describe ➤ Summary Tables.

   Run Graphs from Tasks ➤ Graph.

   Plot the average resolution time and count of observations against the service request status.

Summary Tables

	RESOLUTIONTIME N	RESOLUTIONTIME Mean
ServiceRequestStatus		
Cancel	1	0.00
Closed	1790	56.45

Generated by the SAS System ('SASApp', Linux) on August 18, 2015 at 4:25:36 PM

Remove the observations where Status is Closed because it's an outlier and should not be included in the analysis.

DATA WORK.CS3;
SET WORK.CS2;
IF ServiceRequestStatus='Closed';RUN;

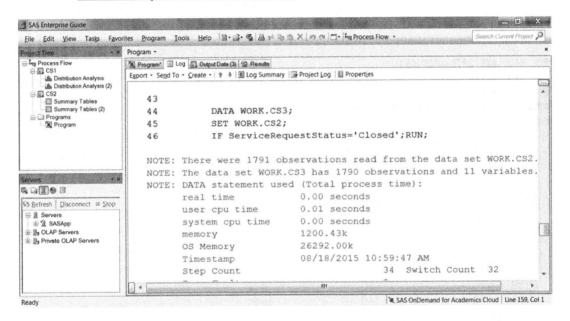

Let's check the spread of Resolution Time over Geography.

■ **Tip**   PROC TABULATE is used to create the summary tables that you can create from the automated button system.

```
PROC TABULATE DATA=WORK.CS3;
 VAR RESOLUTIONTIME;
 CLASS Geography;
TABLE Geography, RESOLUTIONTIME*N RESOLUTIONTIME*MEAN;
RUN;
```

	RESOLUTIONTIME	RESOLUTIONTIME
	N	Mean
Geography		
Asia-Pacific	1790	56.45

▪ **Tip**   A comma is used to separate multiple variables in the code.

Since all the cases belong to the same geography, you can drop this variable.
Resolution Time by Country:

```
PROC TABULATE DATA=WORK.CS3;
 VAR RESOLUTIONTIME;
 CLASS Country;
TABLE Country, RESOLUTIONTIME*N RESOLUTIONTIME*MEAN;
RUN;
```

Page Break

	RESOLUTIONTIME	RESOLUTIONTIME
	N	Mean
Country		
Australia	263	84.19
Banglades	14	125.71
Brunei Da	11	8.00
China	87	54.23
India	746	47.67
Indonesia	50	78.02
Japan	145	55.39
Korea Sou	96	39.31
Malaysia	80	55.18
New Zeala	32	42.38
Oman	3	3.67
Papua New	9	74.11
Philippin	10	52.90
Singapore	164	64.09
Taiwan	17	39.82
Thailand	33	43.52
Unspecifi	17	68.88
Vietnam	13	23.15

Resolution Time by Product:

```
PROC TABULATE DATA=WORK.CS3;
 VAR RESOLUTIONTIME;
 CLASS Product;
TABLE Product, RESOLUTIONTIME*N RESOLUTIONTIME*MEAN;
RUN;
```

Page Break

Product	RESOLUTIONTIME N	RESOLUTIONTIME Mean
AAM	91	53.82
APC	168	77.04
Adva	11	128.82
Blen	71	21.34
Busi	394	50.07
DynA	22	24.27
Oper	280	68.86
PHD	132	82.14
Proc	38	20.47
Prod	3	20.67
USD	503	46.28
Unsp	77	75.00

Page Break

Type Of Engagement for Resolution Time:

Page Break

TypeOfEnagagement	RESOLUTIONTIME N	RESOLUTIONTIME Mean
Billable	43	53.09
Contract	489	56.59
No Charg	351	31.62
Project	901	66.58
Unverifi	6	1.50

Priority level and Resolution Time:

Summary Tables

SR Priority	RESOLUTIONTIME N	RESOLUTIONTIME Sum
	0	.
1-Critic	86	4559.00
2-High	563	27443.00
3-Medium	1006	60403.00
4-Low	136	8644.00

Incident/Problem versus Resolution Time:

	RESOLUTIONTIME N	RESOLUTIONTIME Mean
Incident/ Problem		
Incident	1117	47.54
Problem	497	79.22
Request	176	48.70

11. Create the project datamart. To create the final project datamart, the following data manipulations should be done:

- Remove Geography as a field since all cases are for Asia/Pacific.

- Remove the service request Status as all the statuses are closed.

- Remove SR Open Date as it is used to calculate the *y* variables.

- Remove SR Close Date as it is used to calculate the *y* variable.

- Convert TypeOfEnagagement, Incident..Problem, SR.Priority, Product, Country into numeric fields (dummy and derived variables).

Dropping variables – how will you remove a field.

```
DATA WORK.CS4 (DROP=Geography ServiceRequestStatus 'SR OPEN DATE'n
'sr close date'n);
set work.cs3; RUN;
```

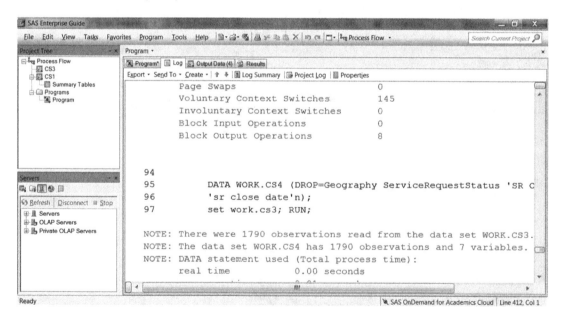

▪ **Tip**   Use an array to create dummy variables in SAS.

A SAS array is a convenient way of temporarily identifying a group of variables for processing within a data step. Once the array has been defined, the same tasks can be performed for a series of related variables, in other words, the array elements.

```
DATA WORK.CS5 (DROP = i);
     SET WORK.CS4;
ARRAY A {*} COUNTRY COUNTRY1-COUNTRY18 ;
DO i = 1 TO 18;
     A(i) = (COUNTRY=i);
END;

PROC FREQ DATA=CS5;
     TABLE COUNTRY1-COUNTRY18;
RUN;
```

Tasks for you:-
a. Create the Dummy Variables for the rest of the variables
b. For SR Priority field, use the code

```
DATA WORK.NewDataFileName;
SET WORK.OldDataFileName;
PRIORITY = SUBSTRN('SR PRIORITY'n,1,1);
RUN;
```

c. Drop the variables from which you have created the Dummy Variables by using code
```
DATA WORK.NewDataFileName (Drop= Var1 var2 .... Varn);
SET WORK.OldDataFileName;
RUN;
```

d. Save the Project Datamart on the Desktop using the File > Export option

Case Study 1 with R

Let's look at how you can solve Case Study 1 using R.

1. Set the path to the directory that contains the Case Study 1 data.

    setwd("H:/springer book/Case study")

    Note that a forward slash or two backward slashes can be used to set the path of the directory or file in R.

2. Import the data.

    cs1 <- read.table("H:/springer book/Case study/CaseStudy1.csv", header=TRUE, sep=",")

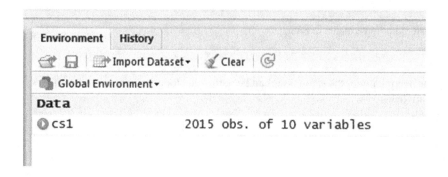

■ **Tip** An alternate way to import the data is to use Tools ➤ Import Data. This will open an Import Wizard, which you can use to bring the data into R.

---

3. View the first ten lines of data to see how the data set looks.

head(cs1, n=10)

```
   ServiceRequestNo ServiceRequestStatus TypeOfEnagagement Incident..Problem
1              1001               Closed           Project          Incident
2              1002               Closed           Project          Incident
3              1003               Closed          Contract           Problem
4              1004               Closed          Contract          Incident
5              1005               Closed          Contract           Problem
6              1006               Closed          Contract           Problem
7              1007               Closed          Contract          Incident
8              1008               Closed          Contract          Incident
9              1009               Closed          Contract          Incident
10             1010               Closed          Contract          Incident
   SR.Priority SR.Open.Date SR.Close.Date Product    Geography   Country
1       2-High     2-Jan-14       1-Mar-14     APC Asia-Pacific  Thailand
2       2-High     2-Jan-14       1-Mar-14     APC Asia-Pacific  Thailand
3       2-High    23-Jan-14      22-Oct-14     APC Asia-Pacific     Japan
4     3-Medium    24-Jan-14      24-Mar-14     APC Asia-Pacific Singapore
5     3-Medium    28-Jan-14      23-Apr-14     APC Asia-Pacific     India
6     3-Medium     5-Mar-14      22-Oct-14     APC Asia-Pacific     Japan
7       2-High     5-Mar-14       4-Apr-14     APC Asia-Pacific     Japan
8     3-Medium    25-Mar-14      30-May-14     APC Asia-Pacific Singapore
9       2-High    25-Mar-14       4-Apr-14     APC Asia-Pacific     Japan
10      2-High    25-Mar-14       4-Apr-14     APC Asia-Pacific     Japan
```

■ **Tip** An alternate way to view the data is to double-click the data set created in the Global Environment window. You can see the file in the Script Editor pane on the top-left side.

4. Understand the types of variables in the data set.

4.1 str(cs1)

```
'data.frame':          2015 obs. of  10 variables:
$ ServiceRequestNo    : int   1001 1002 1003 1004 1005 1006 1007 1008 1009 1010 ...
$ ServiceRequestStatus: Factor w/ 6 levels "Canceled","Closed",..: 2 2 2 2 2 2 2 2 2 2 ...
$ TypeOfEnagagement   : Factor w/ 5 levels "Billable","Contract",..: 4 4 2 2 2 2 2 2 2 2
                        ...
$ Incident..Problem   : Factor w/ 3 levels "Incident","Problem",..: 1 1 2 1 2 2 1 1 1 1 ...
$ SR.Priority         : Factor w/ 4 levels "1-Critical","2-High",..: 2 2 2 3 3 3 2 3 2 2
                        ...
$ SR.Open.Date        : Factor w/ 368 levels "1-Apr-14","1-Apr-15",..: 142 142 190 204 253
                        315 315 219 219 219 ...
$ SR.Close.Date       : Factor w/ 304 levels "","1-Apr-15",..: 7 7 143 161 145 143 240 232
                        240 240 ...
$ Product             : Factor w/ 12 levels "AAM ","Adva",..: 3 3 3 3 3 3 3 3 3 3 ...
$ Geography           : Factor w/ 1 level "Asia-Pacific": 1 1 1 1 1 1 1 1 1 1 ...
$ Country             : Factor w/ 18 levels "Australia","Bangladesh",..: 16 16 7 14 5 7 7
                        14 7 7 ...
```

■ **Note**   A data set is called a *data frame* in R.

■ **Tip**   Visit these web sites for more information about R coding.

www.cookbook-r.com/Data_input_and_output/Loading_data_from_a_file

www.statmethods.net/input/importingdata.html

4.2 Dimensions - Observations and Variables

dim(cs1)

[1] 2015   10

4.3 Names of Variables

names(cs1)
```
[1] "ServiceRequestNo"      "ServiceRequestStatus"  "TypeOfEnagagement"
[4] "Incident..Problem"     "SR.Priority"           "SR.Open.Date"
[7] "SR.Close.Date"         "Product"               "Geography"
[10] "Country"
```

5. Save a copy of the original data (cs1).

   copysc1<-cs1

🐢 Global Environment▾

Data

```
⊙ copysc1           2015 obs. of 10 variables
⊙ cs1               2015 obs. of 10 variables
```

■ **Note**   You can use = or < - interchangeably in the R code.

■ **Tip**   Auto-completion of code in R can be done by using the Tab key if you write the code in the Script Editor pane and *not* in the Console area.

■ **Tip**   For help with code in the Console area, type **help(name of function)** and press Enter. This will open Help in the bottom-left pane.

■ **Tip**   Click the Help tab to type in your query and get help.

---

6. Create a *y* variable of Resolution Time = the difference between SR Close Date and SR Open Date.

   > cs1$ResolutionTime<-cs1$SR.Close.Date-cs1$SR.Open.Date

   ```
   Warning message:
   In Ops.factor(cs1$SR.Close.Date, cs1$SR.Open.Date) :
     '-' not meaningful for factors
   ```

■ **FAQ**   What does this error mean? It means that the format of the variable is a *factor*. Factors are categorical variables that act like dummy variables that R codes for you.

What is the solution? Let's change the format of SR.Close.Date and SR.Open.Date to numbers. OR lets you convert these two variables into numbers and re-import the data.

```
> cs1$SR.Close.Date<-as.Date(cs1$SR.Close.Date ,"%m/%d/%Y" )
> cs1$SR.Open.Date<-as.Date(cs1$SR.Open.Date ,"%m/%d/%Y" )
> str(cs1)
'data.frame':          2015 obs. of  11 variables:
 $ ServiceRequestNo    : int   1001 1002 1003 1004 1005 1006 1007 1008 1009 1010 ...
 $ ServiceRequestStatus: Factor w/ 6 levels "Canceled","Closed",..: 2 2 2 2 2 2 2 2 2 2 ...
 $ TypeOfEnagagement   : Factor w/ 5 levels "Billable","Contract",..: 4 4 2 2 2 2 2 2 2 2
                         ...
 $ Incident..Problem   : Factor w/ 3 levels "Incident","Problem",..: 1 1 2 1 2 2 1 1 1 1 ...
 $ SR.Priority         : Factor w/ 4 levels "1-Critical","2-High",..: 2 2 2 3 3 3 2 3 2 2
                         ...
 $ SR.Open.Date        : Date, format: NA NA ...
 $ SR.Close.Date       : Date, format: NA NA ...
 $ Product             : Factor w/ 12 levels "AAM ","Adva",..: 3 3 3 3 3 3 3 3 3 3 ...
 $ Geography           : Factor w/ 1 level "Asia-Pacific": 1 1 1 1 1 1 1 1 1 1 ...
 $ Country             : Factor w/ 18 levels "Australia","Bangladesh",..: 16 16 7 14 5 7 7
                         14 7 7 ...
 $ ResolutionTime      : logi  NA NA NA NA NA NA ...
```

DATA TYPES IN R

The data types in R are as follows:

- *Numeric*: Decimal numbers.

- *Integer*: Numbers without a fraction/decimal.

- *Complex*: A complex number is a number that can be expressed in the form $a + bi$, where a and b are real numbers and i is the imaginary unit, that satisfies the equation i2 = −1.

  - *Logical*: True/false.

  - *Character*: String values.

- *Vector*: A vector is a sequence of data elements of the same basic data type, generally numbers. Members in a vector are officially called *components*.

- *Matrix*: A matrix is a collection of data elements arranged in a two-dimensional rectangular layout.

- *List.* A list is a generic vector containing numeric, non-numeric, and other objects.

- *Data Frame.* A data frame is a table, or two-dimensional array-like structure, in which each column contains measurements on one variable and each row contains one case.

Note: A data frame is the data table of Excel.

1. Re-import the data. I formatted the two columns of SR.Open.Date and SR.Close. Date as numbers.

```
> setwd("H:/springer book/Case study")
> cs1 <- read.table("H:/springer book/Case study/CaseStudy1.csv", header=TRUE, se
p=",",stringsAsFactors=FALSE)
> str(cs1)
'data.frame':        2015 obs. of  10 variables:
 $ ServiceRequestNo   : int  2090 2863 1517 2864 2865 2866 2022 2061
                            2954 2996 ...
 $ ServiceRequestStatus: chr  "Closed" "Closed" "Closed" "Closed" ...
 $ TypeOfEnagagement  : chr  "Project" "Contract" "Project"
                            "No Charge Support" ...
 $ Incident..Problem  : chr  "Incident" "Problem" "Problem" "Request for
                            Fulfillment" ...
 $ SR.Priority        : chr  "3-Medium" "3-Medium" "3-Medium" "3-Medium" ...
 $ SR.Open.Date       : num  41647 41655 41649 41666 41666 ...
 $ SR.Close.Date      : num  41648 41659 41663 41667 41667 ...
 $ Product            : chr  "Oper" "Busi" "AAM " "Busi" ...
 $ Geography          : chr  "Asia-Pacific" "Asia-Pacific" "Asia-Pacific"
                            "Asia-Pacific" ...
 $ Country            : chr  "Australia" "India" "New Zealand" "India" ...
```

```
> cs1$ResolutionTime<-cs1$SR.Close.Date-cs1$SR.Open.Date
```

```
> View(cs1$ResolutionTime)
```

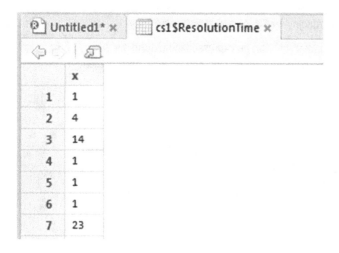

The difference in days has been computed in the resolution time.

2. Understand the distribution of the *y* variable.

```
> d <- density(na.omit(cs1$ResolutionTime))
> plot(d)
```

■ **FAQ** na.omit helps to omit missing values.

---

density.default(x = na.omit(cs1$Resolution Time))

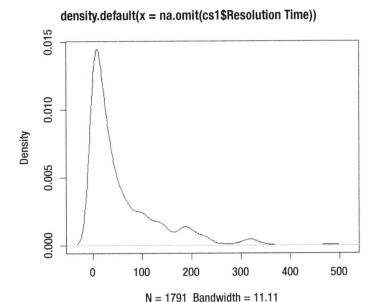

N = 1791 Bandwidth = 11.11

3. View the univariate analysis of the *y* variable. What does the descriptive stats say?

```
> install.packages('pastecs')
```

```
Installing package into 'C:/Users/Subhashini/Documents/R/win-library/3.1'
(as 'lib' is unspecified)
trying URL 'http://cran.rstudio.com/bin/windows/contrib/3.1/pastecs_1.3-18.zip'
Content type 'application/zip' length 1636168 bytes (1.6 MB)
opened URL
downloaded 1.6 MB

package 'pastecs' successfully unpacked and MD5 sums checked
```

The downloaded binary packages are here:

```
C:\Users\Subhashini\AppData\Local\Temp\RtmpuqEEvG\downloaded_packages
```

```
> library(pastecs)
Loading required package: boot

> stat.desc(cs1$ResolutionTime)
        nbr.val      nbr.null       nbr.na           min          max        range
  1.791000e+03  1.720000e+02  2.240000e+02  0.000000e+00  4.840000e+02  4.840000e+02
            sum        median         mean       SE.mean  CI.mean.0.95          var
  1.010490e+05  2.800000e+01  5.642044e+01  1.679654e+00  3.294289e+00  5.052837e+03
        std.dev      coef.var
  7.108331e+01  1.259886e+00
```

Note   To change the output to an easily understood format, change the format of the display.

Hint   Packages in R are a bunch of code written to simplify certain statistical processes or create visualizations and outputs in certain formats.

To see all the statistics, set the following options:

```
> options(scipen=100)
> options(digits=2)
```

Now run the descriptive stats.

```
> stat.desc(cs1$ResolutionTime)
   nbr.val      nbr.null      nbr.na        min          max        range
    1791.0         172.0       224.0        0.0        484.0        484.0
       sum        median        mean     SE.mean  CI.mean.0.95          var
  101049.0          28.0        56.4        1.7          3.3       5052.8
   std.dev      coef.var
      71.1           1.3
```

If you only want the descriptive statistics, such as the min, max, and std.dev, you can add the following option:

```
> stat.desc(cs1$ResolutionTime, basic=F)
   median          mean     SE.mean  CI.mean.0.95          var      std.dev
     28.0          56.4         1.7           3.3       5052.8         71.1
 coef.var
      1.3
```

If you want only the basic statistics such as the number of observations and the number of missing values, use this:

```
> stat.desc(cs1$ResolutionTime, desc=F)
nbr.val nbr.null   nbr.na      min      max    range      sum
   1791      172      224        0      484      484   101049
```

4. Drop the observations with *y* missing.

```
Nbr.null = 172
Nbr.na = 224 (which contains the Nbr.null = 172)
Therefore, total number of observations to be dropped is Nbr.na = 224
```

> **cs2<-cs1[!is.na(cs1$ResolutionTime),]**

Data	
● cs1	2015 obs. of 11 variables
● cs2	1791 obs. of 11 variables

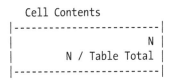 **Tip** Why remove observations with NA? Since *y* is the independent variable and you have enough data (more than 500 observations), you can choose to use the observations that have *y* present.

---

5. Do some visualizations to understand the data.

6. Install a new set of packages to help create the tables and charts.

> **install.packages('gmodels')**

```
Installing package into 'C:/Users/Subhashini/Documents/R/win-library/3.1'
(as 'lib' is unspecified)
trying URL 'http://cran.rstudio.com/bin/windows/contrib/3.1/gmodels_2.16.2.zip'
Content type 'application/zip' length 73922 bytes (72 KB)
opened URL
downloaded 72 KB
```

```
package 'gmodels' successfully unpacked and MD5 sums checked
```

The downloaded binary packages are here:

```
C:\Users\Subhashini\AppData\Local\Temp\RtmpuqEEvG\downloaded_packages
```

> **library(gmodels)**

> **CrossTable(cs2$ServiceRequestStatus)**

```
   Cell Contents
|-------------------------|
|                       N |
|         N / Table Total |
|-------------------------|
```

```
Total Observations in Table:  1791

        | Canceled |   Closed |
        |----------|----------|
        |        1 |     1790 |
        |    0.001 |    0.999 |
        |----------|----------|
```

You need to drop the case that has the status as Canceled. Why? Because there is only one such case and it's obviously an outlier. Also, Resolution Time is measured only for cases that are closed.

```
> cs3 <- cs2[ which(cs2$ServiceRequestStatus=='Closed'), ]
```

Global Environment ▾

Data

▶ cs1	2015 obs. of 11 variables
▶ cs2	1791 obs. of 11 variables
▶ cs3	1790 obs. of 11 variables

Frequency table for TypeOfEnagagement

```
> mytable <- table(cs3$TypeOfEnagagement)

> mytable

            Billable      Contract No Charge Support          Project
                  43           489              351              901
          Unverified
                   6
```

Install libraries. (For packages that are already installed in the system, you need to just use the library function to start using the package.)

```
> library(Lahman)
> library(plyr)

> Table1 <- ddply(cs3, "Incident..Problem", summarise, total = mean(ResolutionTime))

> Table1
        Incident..Problem total
1                 Incident    48
2                  Problem    79
3 Request for Fulfillment    49
```

```
> Table2 <- ddply(cs3, "Product", summarise, total = mean(ResolutionTime))
```

```
> Table2
   Product total
1      AAM    54
2     Adva   129
3      APC    77
4     Blen    21
5     Busi    50
6     DynA    24
7     Oper    69
8      PHD    82
9     Proc    20
10    Prod    21
11    Unsp    75
12     USD    46
```

Sort table Table2 to understand the contribution of each product.

■ **Note**   The default sorting order is ascending. To sort in descending order, add a - (minus) sign in front of the variable.

```
#sort by var1 (ascending) and var2 (descending)
> newdata <- olddata[order(var1, -var2),]
```

```
> names(Table2)
[1] "Product" "total"
```

```
> Table2a <- Table2[order(Table2$total),]
```

```
> Table2a
   Product total
9     Proc    20
10    Prod    21
4     Blen    21
6     DynA    24
12     USD    46
5     Busi    50
1      AAM    54
7     Oper    69
11    Unsp    75
3      APC    77
8      PHD    82
2     Adva   129
```

```
> Table3 <- ddply(cs3, "Geography", summarise, total = mean(ResolutionTime))
```

```
> Table3
      Geography total
1 Asia-Pacific    56
```

■ **Tip** All the cases considered are from the same geography. This field is thus redundant since there is no variation.

Create a table to understand the Country variable.

```
> Table4 <- ddply(cs3, "Country", summarise, total = mean(ResolutionTime))
```

```
> Table4
              Country total
1            Australia  84.2
2           Bangladesh 125.7
3    Brunei Darussalam   8.0
4                China  54.2
5                India  47.7
6            Indonesia  78.0
7                Japan  55.4
8          Korea South  39.3
9             Malaysia  55.2
10         New Zealand  42.4
11                Oman   3.7
12    Papua New Guinea  74.1
13         Philippines  52.9
14           Singapore  64.1
15              Taiwan  39.8
16            Thailand  43.5
17         Unspecified  68.9
18             Vietnam  23.2
```

Let's sort this table in descending order of total.

```
> Table4a <- Table4[order(-Table2$total),]
```

```
> Table4a
              Country total
2           Bangladesh 125.7
8          Korea South  39.3
3    Brunei Darussalam   8.0
11                Oman   3.7
7                Japan  55.4
1            Australia  84.2
5                India  47.7
12    Papua New Guinea  74.1
6            Indonesia  78.0
```

```
4              China  54.2
10       New Zealand  42.4
9           Malaysia  55.2
```

You can see a lot of variation in the average time taken to resolve requests across different countries. Create a table to explore Average Resolution Time across Service Priority

> Table5 <- ddply(cs3, "SR.Priority", summarise, total = mean(ResolutionTime))

> Table5
```
  SR.Priority total
1  1-Critical    53
2      2-High    49
3    3-Medium    60
4       4-Low    64
```

This table is interesting. The Average Resolution Time for High Priority is less than for Critical.

7. To create the final project datamart, the following data manipulations should be done:

   - Remove Geography as a field since all cases are for Asia/Pacific.

   - Remove the service request Status as all statuses are closed.

   - Remove SR Open Date as it is used to calculate the y variable.

   - Remove SR Close Date as it is used to calculate the y variable.

   - Convert TypeOfEnagagement, Incident..Problem, SR.Priority, Product, Country into numeric fields (dummy and derived variables).

> cs3$ServiceRequestStatus<-NULL

⊙ cs3   1790 obs. of 10 variables

> cs3$SR.Open.Date<-NULL

⊙ cs3   1790 obs. of 9 variables

> cs3$SR.Close.Date<-NULL

⊙ cs3   1790 obs. of 8 variables

> cs3$Geography<-NULL

⊙ cs3   1790 obs. of 7 variables

Create dummy variables in the cs3 data frame.

```
> for(level in unique(cs3$TypeOfEnagagement)){cs3[paste("dummy", level, sep = "_")] <-
ifelse(cs3$TypeOfEnagagement == level, 1, 0)}
```

```
> View(cs3)
```

Incident..Problem	SR.Priority	Product	Country	ResolutionTime	dummy_Project	dummy_Contract
Request for Fulfillment	3-Medium	Busi	India	1	0	
Incident	3-Medium	Blen	India	23	1	
Incident	3-Medium	AAM	Singapore	7	0	

```
> for(level in unique(cs3$Incident..Problem)){cs3[paste("dummy", level, sep = "_")] <-
ifelse(cs3$Incident..Problem == level, 1, 0)}
```

```
> for(level in unique(cs3$Product)){cs3[paste("dummy", level, sep = "_")] <-
ifelse(cs3$Product == level, 1, 0)}
```

```
> for(level in unique(cs3$Country)){cs3[paste("dummy", level, sep = "_")] <-
ifelse(cs3$Country == level, 1, 0)}
```

For SR.Priority, you need to extract the first character of the string.

```
> cs3$Priority<- substring(cs3$SR.Priority, 1, 1)
```

(syntax = substr(x, start, stop))

Let's now view the variables in the cs3 data frame, which will serve as the project datamart going forward.

```
> names (cs3)
 [1] "ServiceRequestNo"            "TypeOfEnagagement"
 [3] "Incident..Problem"           "SR.Priority"
 [5] "Product"                     "Country"
 [7] "ResolutionTime"              "dummy_Project"
 [9] "dummy_Contract"              "dummy_No Charge Support"
[11] "dummy_Billable"              "dummy_Unverified"
[13] "dummy_Incident"              "dummy_Problem"
[15] "dummy_Request for Fulfillment" "dummy_Oper"
[17] "dummy_Busi"                  "dummy_AAM "
[19] "dummy_Blen"                  "dummy_Unsp"
[21] "dummy_PHD "                  "dummy_APC "
[23] "dummy_USD "                  "dummy_Adva"
[25] "dummy_Proc"                  "dummy_DynA"
[27] "dummy_Prod"                  "dummy_Australia"
[29] "dummy_India"                 "dummy_New Zealand"
[31] "dummy_Singapore"             "dummy_Japan"
```

```
[33] "dummy_Korea South"            "dummy_Thailand"
[35] "dummy_Taiwan"                 "dummy_Malaysia"
[37] "dummy_Indonesia"             "dummy_Vietnam"
[39] "dummy_Philippines"           "dummy_Bangladesh"
[41] "dummy_Unspecified"            "dummy_China"
[43] "dummy_Brunei Darussalam"      "dummy_Papua New Guinea"
[45] "dummy_Oman"                   "Priority"
```

Let's now drop the original variables on which the dummy and derived variables have been created.

> **cs4<-subset(cs3, , -c(2:6))**

```
cs3                   1790 obs. of 46 variables
cs4                   1790 obs. of 41 variables
```

Now save the project datamart (cs4 file) and other files for later use.

> **save(cs3,file="H:/springer book/Case study/cs3.Rda")**

> **save(cs4,file="H:/springer book/Case study/cs4.Rda")**

■ **Note**   You can load R data files using the load("path/data.Rda") command.

■ ■ ■

## CHAPTER 4

Discover Basic Information About Data Using SAS and R

In this chapter, you will discover basic information about data and learn about descriptive statistics, measures of central tendency, and measures of variation. To fulfill the practical aspect of this chapter, you'll generate this information using SAS and R.

What Are Descriptive Statistics?

The discipline of descriptive statistics is the analysis of data that helps describe or summarize data in a meaningful way. Because of the summarization, some patterns may emerge from the data. The point to note is that any conclusions are made only to the extent or with respect to the data provided and used. The discipline is also not very effective if you need to explore any hypotheses/suppositions you have.

Thus, descriptive statistics describe data and give you a sense of the vital statistics related to the data. It is a good way to summarize row- and column-level data to see how the data looks. Descriptive statistics are used to describe the basic features of the data in a study through simple summaries about the sample and the measures.

Generally, there are two types of statistics that are used to describe data.

- *Measures of central tendency*: These are ways of describing the central position of a frequency distribution or table of data. In other words, how can you understand which is the most frequent or common value? The statistics used include mode, median mean, and so on.

- *Measures of spread or dispersion*: When you look at the data in a spreadsheet or table format with millions of rows, you want to check how the data is spread out, or *dispersed*. This is a way of summarizing a group of data by describing how spread out the data is. The statistics used include range, quartiles, deviation, variance, and standard deviation. Thus, these statistics show you the spread, or *dispersion*.

In plain English, a *distribution* gives you a sense of how things are accumulated in groups. For example, if you have many coins, you will want to create groups of Rs1, Rs2, and Rs5 coins. You will take individual coins and distribute them or stack them in the relevant group. A distribution is a summary of the frequency/count of individual values or ranges of values for a variable. For the variables (generally columns of data), this means that all the data values (rows) may be the presented or grouped into categories first, and then frequencies for each category are calculated.

© Subhashini Sharma Tripathi 2016

S. S. Tripathi, *Learn Business Analytics in Six Steps Using SAS and R*, DOI 10.1007/978-1-4842-1001-7_4

Descriptive statistics are functions of the sample data, and they invariably try to give analysts a sense of how the data looks. A *sample*, as the word denotes, is a small part of the whole data, which is called the *population*. Usually you work on one month or a few months of data. Thus, it is always called a sample. It is often part of *univariate* (one-variable) analysis. As mentioned earlier, these descriptive statistics include mean, minimum, maximum, standard deviation, median, skewness, and kurtosis.

Inferential statistics are assessed as analysts draw an inference regarding a hypothesis about a population parameter. *Inference* in English is a conclusion. You often need some prior understanding of an event to infer with any confidence of being right. Inferential statistics often include z-tests, t-tests, and so on. With inferential statistics, you are trying to reach conclusions that extend beyond the immediate data. These concepts will be covered in later chapters.

More About Inferential and Descriptive Statistics

When you talk about any statistic, be it inferential or descriptive, are you talking about the entire population, in other words, all the data the organization can possibly have?

It is important to note that any statistic, inferential or descriptive, is a function of the sample data. In other words, you will use it after you look up a few rows of the data provided. The status of a given function of the data as a descriptive or inferential statistic depends on the purpose for which the analyst is using it.

When you look at a sample cup of rice (the sample) from a big sack of rice (the population), it is assumed that if the sample has some attribute such as size of grain, the same attribute will be present for the remaining rice in the sack.

Hence, in descriptive statistics, you estimate the population parameters. For example, the sample mean and sample standard deviation provide estimates of the equivalent population parameters. The assumption is that you have picked up data that is a sample that "adequately" represents the population.

How is *adequately* defined? The default definition of this is with "95 percent confidence." Thus, if with 95 percent confidence you can claim that the sample is similar to the population, you are good to go forward with the analysis. (Let's park this word *confidence* on the side for now; we will revisit it in a later chapter.)

Even descriptive statistics such as the minimum and maximum provide information about similar population parameters.

These are two general ways of representing a descriptive statistic:

- Tables

- Graphs

Tables and Descriptive Statistics

An ordered array arranges the values of a numeric variable in rank order from the smallest value to the largest value. This helps an analyst to get a better sense of the range of values that exist in the data. See Figure 4-1.

S. no.	No. of transactions of 10 customers
1	201
2	210
3	219
4	228
5	237
6	246
7	255
8	264
9	273
10	282

Figure 4-1. *Ordered array*

When you have a data set that contains a large number of values, creating an ordered array can be difficult. You may need to divide the data into bands, called *strata segments*. Thus, if you want to create an ordered array for the ages of all the students in the school of 10,000 students, you would create age bands such as less than 6 years old, 6 to 10 years, 10 to 14 years, and older than 14 years. See Figure 4-2.

When you create an ordered array, you can use the following:

- Frequency or percentage distribution

- Cumulative frequency or percentage distribution

What Is a Frequency Distribution?

A *frequency distribution* summarizes numerical values by tallying them into a set of numerically ordered classes/groups. Classes are groups or segments and also called the *class interval*. Frequency is the count of articles that you are dealing with. Each data point, or *observation*, can belong to one class only, as shown in Figure 4-2 earlier. Each student can be counted only once—and only in the Age (class) that the student belongs to. See Figure 4-3.

School with 10,000 students	
Age (in years)	Count of students
<6	1500
6--10	2000
10-14	2500
>14	4000
Total	10000

Figure 4-2. *Frequency distribution*

School with 10,000 students				
Age (in years)	Count of students	Percentage	Cumlative Frequency	Relative Frequency
<6	1500	15%	15%	0.15
6--10	2000	20%	35%	0.20
10--14	2500	25%	60%	0.25
>14	4000	40%	100%	0.40
Total	10000	100%		

Figure 4-3. *Relative and cumulative frequency*

Relative frequency is the frequency of each class divided by the sample size.

A *cumulative frequency distribution* is a summary of data frequency below a given level as a proportion of the sample size.

A cumulative frequency graph is a line graph called an *ogive*. It is plotted showing the cumulative frequency distribution. See Figure 4-4.

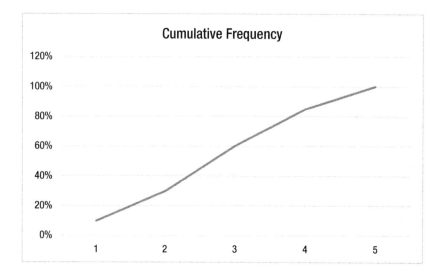

Figure 4-4. *Ogive, looks like an "S"*

Types of data can be divided into the following:

- *Numeric data*: This is data as a measurement, such as weight, height, or count of pages in a book. Often this type of data can have decimal values.

- *Categorical data*: This data represents characteristics such as gender, status, types of movies, and names of cities. Categorical data can have numerical values by it. These numbers don't have any mathematical meaning. Qualitative data or yes/no data will also come into play here.

  - *Discrete data (numeric)*: This represents items that can be counted in whole numbers. An example is the count of students in a class or the number of cars. You cannot have any decimal values here.

- *Continuous data (numeric)*: This represents items that can be counted in fractions as well as in whole numbers. Age and weight are examples. You can have decimal values here.

■ **Note**   Ordinal data mixes numerical and categorical data. The data falls into categories, but the numbers used to represent these categories have meaning. For example, customer service feedback can range from 0 to 4 stars, with 0 being the least and 4 being the highest.

When you organize the data, you sometimes begin to discover patterns on relationships in the data. However, the type of visual representation will often depend on the following:

- The type of data

- The purpose or question you are seeking to answer

    - *Bar charts/column charts (vertical)*: These are best for comparing counts, sums, or percentages of two to seven groups. Each bar is separated by a blank space. The bar chart should be used only to compare categories that are mutually exclusive.

    - *Bar charts (horizontal)*: These are used for comparing numbers across eight or more groups. All the categories should be measured independently.

    - *Pie charts*: This is used to show different groups/competence of a single dimension. A pie chart should be used only within a group of categories that combine to make up a whole, or 100 percent.

    - *Line charts*: These illustrate trends over periods of time and are used to measure the long-term progression of statistics. Multiple variables can be plotted as long as the time remains the same.

    - *Scatter plot*: These are used to depict how different objects/measurements of variables settle on two or three different dimensions. These are frequently used in analytics.

    - *Histogram*: Like pie charts, histograms break down the sample distribution in one dimension. They are bar charts without any space between the bars. These are often used to depict the distribution of variables.

Case Study 2

Xenon is an Internet marketing company. It has data related to the marketing revenues of its customers. Most of its customers plan their marketing budgets on a monthly basis. Xenon wants to understand its data better. It is looking at creating some tables and charts to get some insight into its data.

The components of its data are as follows:

- *Year*: All the data belongs to the year 2014.

- *Month*: This is the month for which the data belongs.

- *Quarter*: The quarter is Q1 to Q3 2014.

- *Customer type*: Is it a new customer or an existing customer for the company?

- *Type of calling*: How did the customer engage with the company? Did the customer call in (inbound) and inquire about or request services? Or did the company call the customer (outbound) and engage with the customer?

- *Vertical*: This is the industry to which the customer belongs.

- *Monthly recurring revenue in INR*: This is the amount of money the customer paid the company in a particular month. All amounts are in Indian rupees.

Here's the general solution process:

1. *Define*: The customer wants to create tables and charts to understand the data better (descriptive statistics).

2. *Collect*: All the data required for this exercise has been given along with an understanding of the fields/variables that exist in the data.

3. *Organize*: Some basic checking for missing values and outliers is all that is required for this exercise.

4. *Visualize*: Tables and graphs have to be created to represent/summarize the data.

5. *Analyze*: No particular statistical analysis is required.

6. *Insights*: Conclusions are to be drawn on the basis of the visualizations.

Solving Case Study 2 with SAS

Prepare the directory/space where you will work. In this code, you tell the system where the data that you will work on is stored.

```
libname lib1 " /home/subhashini1/my_content "; run;
PROC CONTENTS data = lib1._ALL_ NODS;run;
```

The CONTENTS Procedure

Directory	
Libref	LIB1
Engine	V9
Physical Name	/home/subhashini1/my_content
Filename	/home/subhashini1/my_content
Inode Number	10354689
Access Permission	rwxr-xr-x
Owner Name	subhashini1
File Size	4KB
File Size (bytes)	4096

#	Name	Member Type	File Size	Last Modified
1	CS1	DATA	384KB	08/10/2015 10:43:29
2	MARKET1	DATA	256KB	09/26/2015 02:30:26
3	MARKET3	DATA	256KB	09/26/2015 02:30:26

Page Break

```
# Import the data
FILENAME REFFILE "/home/subhashini1/my_content/MarketingData.csv" TERMSTR=CR;
RUN;
/* Comment - I have renamed the file to REFILE so that in the Proc Import step I do not have
to put the full path.*/

PROC IMPORT DATAFILE=REFFILE
    DBMS=CSV
    OUT=WORK.MARKET1;
    GETNAMES=YES;
RUN;
```

The types of delimiters can be the following:

- Comma

- Tab

- Semicolon

- Pipe

- Caret

Understand the dimensions of the data set (called Obs and Vars) and the type of data in the temporary SAS file in the working directory. PROC CONTENTS is one of the most used procedures for the initial understanding of the data size.

```
PROC CONTENTS DATA=WORK.MARKET1; RUN;
```

Page Break

The CONTENTS Procedure

Data Set Name	WORK.MARKET1	Observations	587
Member Type	DATA	Variables	7
Engine	V9	Indexes	0
Created	11/16/2015 17:28:34	Observation Length	72
Last Modified	11/16/2015 17:28:34	Deleted Observations	0
Protection		Compressed	NO
Data Set Type		Sorted	NO
Label			
Data Representation	SOLARIS_X86_64, LINUX_X86_64, ALPHA_TRU64, LINUX_IA64		
Encoding	utf-8 Unicode (UTF-8)		

Alphabetic List of Variables and Attributes				
# Variable	Type	Len	Format	Informat
2 Month	Char	7	$7.	$7.
4 CustomerType	Char	17	$17.	$17.
7 Monthly Reccuring Revenue in INR	Num	8	BEST12.	BEST32.
3 Quarter	Char	2	$2.	$2.
5 TypeOfCalling	Char	8	$8.	$8.
6 Vertical	Char	25	$25.	$25.
1 Year	Char	5	$5.	$5.

Page Break

■ **Note** The PROC CONTENTS procedure gives you details of the data set in one go. If the data variables have labels, explanations, or definitions, they will also be seen in the PROC CONTENTS output.

Save the SAS data set to the permanent directory (the next time you can load this data set instead of the CSV data set).

```
DATA LIB1.MARKET1;
SET WORK.MARKET1; RUN;
```

■ **Note** For steps where no output is desired, you can verify the execution on the Log tab.

The Log tab reads as follows:

```
47          DATA LIB1.MARKET1;
48          SET WORK.MARKET1; RUN;

NOTE: There were 587 observations read from the data set WORK.MARKET1.
NOTE: The data set LIB1.MARKET1 has 587 observations and 7 variables.
NOTE: DATA statement used (Total process time):
      real time            0.05 seconds
      user cpu time        0.01 seconds
# Check to confirm that the file has been saved

PROC CONTENTS data = lib1._ALL_ NODS;run;
```

Page break

The CONTENTS Procedure

	Directory
Libref	LIB1
Engine	V9
Physical Name	/home/subhashini1/my_content
Filename	/home/subhashini1/my_content
Inode Number	10354689
Access Permission	rwxr-xr-x
Owner Name	subhashini1
File Size	4KB
File Size (bytes)	4096

#	Name	Member Type	File Size	Last Modified
1	CS1	DATA	384KB	08/10/2015 10:43:29
2	MARKET1	DATA	256KB	11/16/2015 12:14:12
3	MARKET3	DATA	256KB	09/26/2015 02:30:26

```
# Summary of Recurring Revenue

PROC UNIVARIATE DATA=WORK.MARKET1 ;
VAR 'Monthly Reccuring Revenue in INR'n; RUN;
```

■ **Note** The n at the end of the variable name ensures that the blank spaces in the variable name are overlooked by the SAS system. As you will remember, the SAS system does not understand blanks and any other special characters such as % : ; and so on. It understands only the underscore (_) in place of blanks and other special characters.

The UNIVARIATE Procedure

This section explains the output of the UNIVARIATE procedure. The output I'll discuss is as follows:

Variable: Monthly Recurring Revenue in INR

Moments

N	586	Sum Weights	586	
Mean	347768.848	Sum Observations	203792545	
Std Deviation	2701402.67	Variance	7.30E+12	
Skewness	13.6490561	Kurtosis	207.858855	
Uncorrected SS	4.34E+15	Corrected SS	4.27E+15	
Coeff Variation	776.781097	Std Error Mean	111593.916	

Basic Statistical Measures

Location		Variability	
Mean	347768.8	Std Deviation	2701403
Median	4506	Variance	7.30E+12
Mode	0	Range	46132500
		Interquartile Range	12955

Tests for Location: Mu0=0

Test	Statistic		p Value	
Student's t	t	3.116378	Pr > |t|	0.0019
Sign	M	272	Pr >= |M|	<.0001
Signed Rank	S	74120	Pr >= |S|	<.0001

Quantiles (Definition 5)

Level	Quantile
100% Max	46132500
99%	7152825
95%	520200
90%	188340
75% Q3	14500
50% Median	4506

(*continued*)

25% Q1	1545		
10%	500		
5%	0		
1%	0		
0% Min	0		

Extreme Observations

Lowest		Highest	
Value	Obs	Value	Obs
0	576	7789950	316
0	569	13000000	298
0	546	16275000	248
0	545	38000000	63
0	544	46132500	559

Missing Values

Missing Value	Count	Percent Of All Obs	Missing Obs
.	1	0.17	100

- *Inference*: At least 5 percent of the values are 0. These can be dropped from the analysis.

- *Inference*: There is one missing value. This observation can be dropped from the data set.

Now let's go through this output. "Moments" is the header and explains the descriptive statistics (mean, median, mode, range, and so on).

- *N*: This is the count of observations (rows of data under the variable).

- *Mean*: This is the average value of the numeric variable; it is the arithmetic mean value.

- *Std Deviation*: This is the square root of the variance.

- *Skewness*: This is the symmetry. This shows if there are more values on any one side of the mean. If the values are equally divided, then the skewness is 0.

- *Uncorrected SS*: This value is the sum of squares.

- *Coeff Variation*: This is another way to measure variability.

- *Sum Weights*: The default value is 1. If you specify the weight of the variable to be higher (in other words, the variable is more important than the other variables), then the value you have assigned comes here.

- *Sum Observation*: This is the sum of the weight into the value of the variable.

- *Variance*: This is the variability from the mean.

- *Kurtosis*: This shows the heaviness of the tails. In other words, it shows the flatness of the normal distribution hump.

- *Corrected SS*: This is the sum of square of the distance of values from the mean value.

- *Std Error Mean*: This represents the standard deviation of the mean of the sample from the population.

The Basic Descriptive statistics (which describe and tell you some initial information about the data) are in the "Basic Statistical Measure" section.

- *Mean*: This is the arithmetic mean or average.

- *Median*: This is the middle value. You can sort the data in ascending/descending order and then pick the value in the middle.

- *Mode*: This is the most common value.

- *Std Deviation*: This is the square root of the variance.

As you will see, the variance is the square of the distance between the value and the mean. The positive and negative values are represented in total here because when a minus number is squared, it becomes positive.

In the variance, the numbers look large because they're squared.

- *Variance*: Variance is the square of the distance between the value and the mean. The positive and negative values are represented in total here because when a minus number is squared, it become positive.

- *Range*: The minimum and maximum values define the range.

- *Inter Quartile range*: This is the difference between two quartiles. Quartiles are calculated in the same way as median: by sorting the data and choosing the 25 percent, 50 percent, and 75 percent values.

The next section in the "Univariate" output includes the tests for location.

- *Student's t*: This tests the hypothesis that the population mean is 0.

■ **Note**   Hypothesis testing is covered in subsequent chapters.

- *Sign*: This is a hypothesis test for the median.

- *Signed rank*: This is the Wilcoxon test. You will realize that lots of tests are named after the mathematicians who first used them. Often the difference is in the calculation process or methodology. This test is again for the hypothesis that the median is 0.

"Quantiles" is the next part of the output. This is self-explanatory as it gives the values in percentile after sorting out the variable.

The utility of this segment is as follows:

- Finding outliers (the extreme values)

- Grouping or binning the variables since you can see which values can be used to create bins

Now let's look at some code to deal with missing values. As discussed, dealing with missing values is an important part of the organizing section in DCOVA and I.

```
/* Remove Missing value in Monthly Reccuring Reveue */
DATA WORK.MARKET2;
SET WORK.MARKET1;
IF  'Monthly Reccuring Revenue in INR'n = . THEN DELETE; RUN;
Log shows:-
NOTE: There were 587 observations read from the data set WORK.MARKET1.
NOTE: The data set WORK.MARKET2 has 586 observations and 7 variables.

/*Remove 0 in Monthly Reccuring Revenue*/
DATA WORK.MARKET2;
SET WORK.MARKET2;
IF 'Monthly Reccuring Revenue in INR'n = 0 THEN DELETE; RUN;
Log shows:-
NOTE: There were 544 observations read from the data set WORK.MARKET2.
NOTE: The data set WORK.MARKET2 has 544 observations and 7 variables.
# Create Frequency Tables

/*frequency tables*/

PROC SORT DATA = WORK.MARKET2;
BY CustomerType; RUN;
```

■ **Note**   Why do you sort the data before using the frequency? This is to avoid running out of memory while doing the frequency. If all the values of a particular type are close together, the frequency procedure is able to process the data set using less memory.

```
PROC FREQ DATA=WORK.MARKET2;
 TABLES CustomerType; RUN;
```

The FREQ Procedure				
			Cumulative	Cumulative
CustomerType	Frequency	Percent	Frequency	Percent
0	77	14.15	77	14.15
Existing Customer	188	34.56	265	48.71
New Customer	279	51.29	544	100

The options to PROC FREQ are as follows:

```
proc freq ;
by variables ;
exact statistic-options < / computation-options> ;
output options ;
tables requests < /options> ;
```

```
test options ;
weight variable ;
```

These options are described as follows:

- BY calculates separate frequency or cross-tabulation tables for each BY group.

- EXACT requests exact tests for specified statistics.

- OUTPUT creates an output data set that contains the specified statistics.

- TABLES specifies the frequency or cross-tabulation tables and requests tests and measures of association.

- TEST requests asymptotic tests for measures of association and agreement.

- WEIGHT identifies a variable with values that weight each observation.

Let's run it again to see the TABLES option.

```
PROC FREQ DATA=WORK.MARKET2;
TABLES CustomerType*Vertical; RUN;
```

The FREQ Procedure

Table of CustomerType by Vertical

CustomerType		0	Automotive	Business Services	Consumer Goods	Consumer Services	Education	Energy & Utilities	Financial Services	Foundation-Not for Profit	Gaming	High Technology	Hotel & Travel	Manufacturing	Media & Entertainment	Miscellaneous	Not Defined	Pharma Health Care
0	Frequency	0	0	10	2	0	1	0	6	0	8	15	4	4	12	1	0	:
	Percent	0	0	1.84	0.37	0	0.18	0	1.1	0	1.47	2.76	0.74	0.74	2.21	0.18	0	0.18
	Row Pct	0	0	12.99	2.6	0	1.3	0	7.79	0	10.39	19.48	5.19	5.19	15.58	1.3	0	1.:
	Col Pct	0	0	21.74	10.53	0	16.67	0	13.64	0	20.51	23.08	22.22	18.18	10	25	0	12.:
Existing Customer	Frequency	0	3	11	4	0	1	1	20	0	6	20	9	5	39	1	0	:
	Percent	0	0.55	2.02	0.74	0	0.18	0.18	3.68	0	1.1	3.68	1.65	0.92	7.17	0.18	0	0.5:
	Row Pct	0	1.6	5.85	2.13	0	0.53	0.53	10.64	0	3.19	10.64	4.79	2.66	20.74	0.53	0	1.:
	Col Pct	0	75	23.91	21.05	0	16.67	100	45.45	0	15.38	30.77	50	22.73	32.5	25	0	37.:
New Customer	Frequency	2	1	25	13	2	4	0	18	4	25	30	5	13	69	2	1	:
	Percent	0.37	0.18	4.6	2.39	0.37	0.74	0	3.31	0.74	4.6	5.51	0.92	2.39	12.68	0.37	0.18	0.7:
	Row Pct	0.72	0.36	8.96	4.66	0.72	1.43	0	6.45	1.43	8.96	10.75	1.79	4.66	24.73	0.72	0.36	1.4:
	Col Pct	100	25	54.35	68.42	100	66.67	0	40.91	100	64.1	46.15	27.78	59.09	57.5	50	100	5(
Total	Frequency	2	4	46	19	2	6	1	44	4	39	65	18	22	120	4	1	:
	Percent	0.37	0.74	8.46	3.49	0.37	1.1	0.18	8.09	0.74	7.17	11.95	3.31	4.04	22.06	0.74	0.18	1.4:

```
/* the table has too many outputs like Row Pct , Col Pct etc.*/

PROC FREQ DATA=WORK.MARKET2;
TABLES Vertical*CustomerType/norow nocol nopercent; RUN;

# Sort the out the table for easier understanding

PROC FREQ DATA=WORK.MARKET2 ORDER=FREQ;
 TABLES Vertical*CustomerType/norow nocol nopercent; RUN;
```

The FREQ Procedure				
Table of Vertical by CustomerType				
		CustomerType		
		0	Existing Customer	New Customer	Total
Vertical					
0	Frequency	0	0	2	2
Automotive	Frequency	0	3	1	4
Business Services	Frequency	10	11	25	46
Consumer Goods	Frequency	2	4	13	19
Consumer Services	Frequency	0	0	2	2
Education	Frequency	1	1	4	6
Energy & Utilities	Frequency	0	1	0	1
Financial Services	Frequency	6	20	18	44
Foundation-Not for Profit	Frequency	0	0	4	4
Gaming	Frequency	8	6	25	39
High Technology	Frequency	15	20	30	65
Hotel & Travel	Frequency	4	9	5	18
Manufacturing	Frequency	4	5	13	22
Media & Entertainment	Frequency	12	39	69	120
Miscellaneous	Frequency	1	1	2	4
Not Defined	Frequency	0	0	1	1
Pharma/Health Care	Frequency	1	3	4	8
Public Sector	Frequency	1	11	17	29
Retail	Frequency	10	53	33	96
Software as a Service	Frequency	2	1	11	14
Total	Frequency	77	188	279	544

The order = option orders the values of the frequency and cross-tabulation table variables according to the specified order, where:

- data orders values according to their order in the input data set.

- formatted orders values by their formatted values.

- freq orders values by descending frequency count.

- internal orders values by their unformatted values.

The following code is written to create a frequency/count by the segments under the variable:

```
# Frequency table for Verticals

PROC FREQ DATA=WORK.MARKET2 ORDER=FREQ;
TABLES Vertical/norow nocol; RUN;
```

This code creates a table where you can see the following:

- Frequency or count under each

- Percent contribution of each segment to the whole

- Cumulative frequency count

- Cumulative percent, which is the percent of frequency in 100 percent of the segments

Vertical	Frequency	Percent	Cumulative Frequency	Cumulative Percent
Media & Entertainment	120	22.06	120	22.06
Retail	96	17.65	216	39.71
High Technology	65	11.95	281	51.65
Business Services	46	8.46	327	60.11
Financial Services	44	8.09	371	68.20
Gaming	39	7.17	410	75.37
Public Sector	29	5.33	439	80.70
Manufacturing	22	4.04	461	84.74
Consumer Goods	19	3.49	480	88.24
Hotel & Travel	18	3.31	498	91.54
Software as a Service	14	2.57	512	94.12
Pharma/Health Care	8	1.47	520	95.59
Education	6	1.10	526	96.69
Automotive	4	0.74	530	97.43

(continued)

Vertical	Frequency	Percent	Cumulative Frequency	Cumulative Percent
Foundation-Not for Profit	4	0.74	534	98.16
Miscellaneous	4	0.74	538	98.90
0	2	0.37	540	99.26
Consumer Services	2	0.37	542	99.63
Energy & Utilities	1	0.18	543	99.82
Not Defined	1	0.18	544	100.00

There are far too many verticals listed in the data set. Most of them have a less than 5 percent contribution to the data set (shown in the previous table). It will be desirable to club all the verticals except the top four. Create a new variable to do so.

```
DATA WORK.MARKET3;
SET WORK.MARKET2;
IF VERTICAL = "Media & Entertainment" THEN NEWVAR=1;
ELSE IF VERTICAL = "Retail" THEN NEWVAR=2;
ELSE IF VERTICAL = "High Technology" THEN NEWVAR=3;
ELSE IF VERTICAL = "Business Services" THEN NEWVAR=4;
ELSE NEWVAR=0;
RUN;

PROC FREQ DATA=WORK.MARKET3;
TABLES NEWVAR ; RUN;
```

NEWVAR	Frequency	Percent	Cumulative Frequency	Cumulative Percent
0	217	39.89	217	39.89
1	120	22.06	337	61.95
2	96	17.65	433	79.60
3	65	11.95	498	91.54
4	46	8.46	544	100.00

Create some tables and graphs to understand the data with respect to the continuous variable of Recurring Monthly Revenue.

You use the means procedure to get numeric summaries across variables and segments in a variable.

```
PROC MEANS DATA=WORK.MARKET3;
VAR 'Monthly Reccuring Revenue in INR'n; RUN;
```

Analysis Variable : Monthly Reccuring Revenue in INR				
N	Mean	Std Dev	Minimum	Maximum
544	374618.65	2802133.63	53.0000000	46132500.00

■ **Note** The Monthly Recurring Revenue values are in very large numbers. Therefore, you will divide the values by 1,000 to get numbers that you can interpret easily.

Create a new variable to represent Monthly Recurring Revenue / 1000 (in thousands).

```
DATA WORK.MARKET3;
SET WORK.MARKET3;
MRR_THOU= 'Monthly Reccuring Revenue in INR'n/1000; RUN;

PROC MEANS DATA=WORK.MARKET3;
VAR MRR_THOU; RUN;
```

Analysis Variable : MRR_THOU				
N	Mean	Std Dev	Minimum	Maximum
544	374.6186489	2802.13	0.0530000	46132.50

Create plots to explore the sum of the new variable MRR_THOU over the other variables. You are doing this to understand how values change across segments and subsegments.

```
PROC MEANS DATA=WORK.MARKET3;
CLASS CustomerType;
VAR MRR_THOU;RUN;
```

Analysis Variable : MRR_THOU						
CustomerType	N Obs	N	Mean	Std Dev	Minimum	Maximum
0	77	77	244.0481818	752.1983309	0.6370000	4626.50
Existing Customer	188	188	314.8758138	3380.08	0.0530000	46132.50
New Customer	279	279	450.9110466	2735.68	0.3850000	38000.00

```
PROC FREQ DATA=WORK.MARKET3;
TABLES CustomerType /NOROW NOCOL;RUN;
```

We use NROW NCOL option to get a clean table output, without summaries across rows and columns.

CustomerType	Frequency	Percent	Cumulative Frequency	Cumulative Percent
0	77	14.15	77	14.15
Existing Customer	188	34.56	265	48.71
New Customer	279	51.29	544	100.00

- *Takeaway 1*: New customers are giving the highest MRR and form the largest chunk of the business (51 percent).

In the following code, you will customize the PROC MEANS output so that you can see the output in a format of your choosing (instead of the default format).

```
PROC MEANS DATA=WORK.MARKET3 N MEAN SUM;
CLASS CustomerType Quarter ;
VAR MRR_THOU;RUN;
```

The MEANS Procedure

Analysis Variable : MRR_THOU					
Customer Type	Quarter	N Obs	N	Mean	Sum
0	Q2	77	77	244.0481818	18791.71
Existing Customer	Q1	56	56	102.6091429	5746.11
	Q2	98	98	41.4042959	4057.62
	Q3	34	34	1452.73	49392.92
New Customer	Q1	143	143	496.2723846	70966.95
	Q2	98	98	475.6756429	46616.21
	Q3	38	38	216.3425789	8221.02

```
PROC MEANS DATA=WORK.MARKET3 N MEAN SUM;
CLASS Quarter ;
VAR MRR_THOU;RUN;
```

The MEANS Procedure

Analysis Variable : MRR_THOU			
Quarter	N Obs	Mean	Sum
Q1	199	385.4927789	76713.06
Q2	273	254.4525421	69465.54
Q3	72	800.1935833	57613.94

- *Takeaway 2*: The quarter-wise MRR is showing a decreasing trend. The count of new customers is highest in Q1, and this is driving the trend.

- *Takeaway 3*: You can see that the average MRR for new customers is the highest.

- *Assignment*: Do a similar exploration for verticals (through the column newvar) and create some takeaway points.

Solving Case Study 2 with R

Prepare the directory/space where you will work.

```
> setwd("H:/springer book/Case study/CaseStudy2")
> getwd()
[1] "H:/springer book/Case study/CaseStudy2"
```

Import the data.

```
market1 <- read.table("H:/springer book/Case study/CaseStudy2/MarketingData.csv",
header=TRUE, sep=",", stringsAsFactors = FALSE)
```

Common separators include the following:

- Comma

- Tab

- Semicolon

- Pipe

- Caret

Now it's time to understand the dimensions of the data set called Obs and Vars.

Variables are the various fields in the data, and observations are the values for the variable. When you import data that you have never seen, you will want to get a sense of how it looks.

```
> dim(market1)
[1] 586    7
```

You want to find out what type (numeric, character, and so on) of data exists for each variable.

```
> str(market1)
 'data.frame':                    586 obs. of  7 variables:
 $ Year                         : int  2014 2014 2014 2014 2014 2014 2014 2014 2014
                                  2014 ...
 $ Month                        : chr  "January" "January" "January" "January" ...
 $ Quarter                      : chr  "Q1" "Q1" "Q1" "Q1" ...
 $ CustomerType                 : chr  "New Customer" "New Customer" "New Customer"
                                  "New Customer" ...
 $ TypeOfCalling                : chr  "Inbound" "Inbound" "Outbound" "Inbound" ...
 $ Vertical                     : chr  "Media & Entertainment" "Education"
"Manufacturing" "High Technology" ...
 $ Monthly.Reccuring.Revenue.in.INR: int  0 1200 3838 8013 18618 1450 4500 8750 2700 4283
                                  ...
```

You want to look at some basic descriptive statistics to understand the numerical variables.

```
> summary(market1$Monthly.Reccuring.Revenue.in.INR)
   Min.  1st Qu.   Median     Mean  3rd Qu.     Max.
      0     1546     4506   347800    14500 46130000
#  Standard Deviation of Recurring Revenue
> sd(market1$Monthly.Reccuring.Revenue.in.INR)
[1] 2701403
```

In R you have some user-defined code called *packages*. You can use them to reduce the coding requirements.

```
> install.packages("psych")
> library(psych)
```

```
> describe(market1$Monthly.Reccuring.Revenue.in.INR, na.rm = TRUE, interp=FALSE,skew = TRUE,
ranges = TRUE,trim=.1,type=3,check=TRUE)
  vars   n     mean      sd median  trimmed     mad min      max    range  skew kurtosis
1    1 586 347768.8 2701403   4506 14927.69 5333.65   0 46132500 46132500 13.58   205.37
        se
1 111593.9
```

The following are the arguments in the `describe` function:

- `x`: This specifies a data frame or matrix.

- `na.rm`: The default is to delete missing data. `na.rm=FALSE` will delete the case.

- `interp`: This specifies whether the median should be standard or interpolated.

- `skew`: This specifies whether the skew and kurtosis should be calculated.

- `ranges`: This specifies whether the range should be calculated.

- `trim`: `trim=.1` means dropping the top and bottom trim fraction.

- `type`: This specifies which estimate of skew and kurtosis should be used.

- `check`: This specifies whether you should check for non-numeric variables. It's slower but helpful.

You want to look at the count of data in different segments, so you'll create frequency tables, as shown here:

```
> attach(market1)
> table(market1$CustomerType)

         0 Existing Customer    New Customer
        88              191             307

> table(market1$TypeOfCalling)

 Inbound Outbound
     128      458

> table(market1$TypeOfCalling,market1$CustomerType)

          0 Existing Customer New Customer
 Inbound  26               10           92
 Outbound 62              181          215

> table1<- table(market1$TypeOfCalling,market1$CustomerType)

> margin.table(table1,1)

 Inbound Outbound
     128      458
```

■ **Note**   1 in the previous code refers to the variable column.

```
> margin.table(table1,2)

          0 Existing Customer    New Customer
     88               191             307
```

If you want to look at the way the data is distributed within segments, it's easier to look at what percentage of the whole is contributed by each segment.

```
> prop.table(table1) # cell percentages

            0 Existing Customer New Customer
  Inbound  0.04436860      0.01706485   0.15699659
  Outbound 0.10580205      0.30887372   0.36689420

> prop.table(table1, 1)   # row percentages

            0 Existing Customer New Customer
  Inbound  0.2031250       0.0781250    0.7187500
  Outbound 0.1353712       0.3951965    0.4694323

> prop.table(table1,2) # column percentages

            0 Existing Customer New Customer
  Inbound  0.29545455      0.05235602   0.29967427
  Outbound 0.70454545      0.94764398   0.70032573

> table(market1$Vertical)
```

0	Automotive	Business Services
3	4	48
Consumer Goods	Consumer Services	Education
19	2	6
Energy & Utilities	Financial Services	Foundation-Not for Profit
1	45	4
Gaming	High Technology	Hotel & Travel
55	66	18
Manufacturing	Media & Entertainment	Miscellaneous
23	138	4
Not Defined	Pharma/Health Care	Public Sector
1	8	30
Retail	Software as a Service	
97	14	

Sort the table to make sense of it.

```
table2<-table(market1$Vertical)

> table4<-as.data.frame(table2) # convert to data frame

> View(table4)

> table5<- table4[order(-table4$Freq),]

sum(table4$Freq)
[1] 586

> table4$cfp<- table4$Freq/586
> table4
                      Var1 Freq         cfp
1                        0    3 0.005119454
2                Automotive    4 0.006825939
3         Business Services   48 0.081911263
4           Consumer Goods   19 0.032423208
5         Consumer Services    2 0.003412969
6                 Education    6 0.010238908

> table5<- table4[order(-table4$cfp),]
> View(table5)

> table5
                      Var1 Freq         cfp
14    Media & Entertainment  138 0.235494881
19                   Retail   97 0.165529010
11          High Technology   66 0.112627986
10                   Gaming   55 0.093856655
3         Business Services   48 0.081911263
8         Financial Services   45 0.076791809
18            Public Sector   30 0.051194539
13            Manufacturing   23 0.039249147
4           Consumer Goods   19 0.032423208
12            Hotel & Travel   18 0.030716724
20      Software as a Service   14 0.023890785
17       Pharma/Health Care    8 0.013651877
6                 Education    6 0.010238908
2                Automotive    4 0.006825939
9   Foundation-Not for Profit    4 0.006825939
15            Miscellaneous    4 0.006825939
1                        0    3 0.005119454
5         Consumer Services    2 0.003412969
7        Energy & Utilities    1 0.001706485
16              Not Defined    1 0.001706485
```

There are far too many verticals listed in the data set. Most of them have a less than 5 percent contribution to the data set (refer to the previous table). It will be desirable to club all the verticals except the top four. Create a new variable to do so.

```
> market1$newvar[market1$Vertical=="Media & Entertainment"]<-1
> View(market1)
> market1$newvar[market1$Vertical=="Retail"]<-2
> market1$newvar[market1$Vertical=="High Technology"]<-3
> market1$newvar[market1$Vertical=="Gaming"]<-4

> market1[is.na(market1)]<-0 .
> table(market1$newvar)

   0   1   2   3   4
 230 138  97  66  55
```

Create some tables and graphs to understand the data with respect to the continuous variable of Recurring Monthly Revenue.

```
library("ggplot2")
```

```
> library("ggplot2")
> qplot(newvar,Monthly.Reccuring.Revenue.in.INR, data = market1)
```

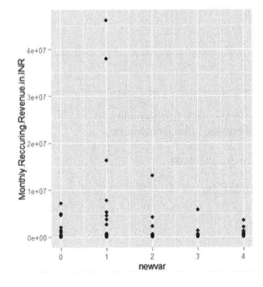

The Monthly Recurring Revenue values are large numbers and appear as exponential.

```
> market1$TypeOfCalling <- as.factor(market1$TypeOfCalling)
> qplot(newvar,Monthly.Reccuring.Revenue.in.INR, data = market1, color=TypeOfCalling)
```

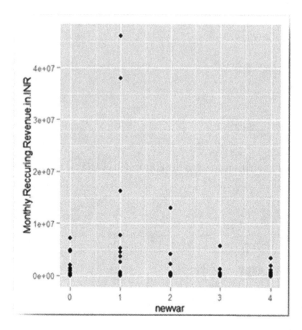

■ **Note**   Include the type of calling as a variable in the graph.

Create a new variable to represent Monthly Recurring Revenue / 1000 (in thousands).
It is easier to understand and represent large numbers when seen in smaller chunks.

```
> market1$MRR.thou<-market1$Monthly.Reccuring.Revenue.in.INR/1000
```

```
> qplot(newvar,MRR.thou, data = market1, color=TypeOfCalling)
```

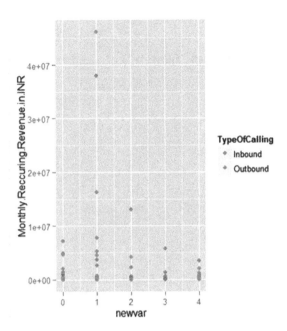

You want to explore which fields are present in the data.

```
> names(market1)
[1] "Year"                         "Month"
[3] "Quarter"                      "CustomerType"
[5] "TypeOfCalling"                "Vertical"
[7] "Monthly.Reccuring.Revenue.in.INR" "newvar"
[9] "MRR.thou"
```

Remove variables where Monthly Recurring Revenue is 0.
These are missing values; hence, you should remove rows with 0 as a value.

```
> market2<- market1[which( market1$Monthly.Reccuring.Revenue.in.INR>0),]

> dim(market2)
[1] 544    9
```

Let's see how to create some tables and graphs to get some visual understanding of the data.

```
> ggplot(market2, aes(x=CustomerType, y=MRR.thou)) + geom_bar(stat="identity")

> table1<- table(market2$CustomerType)

> prop.table(table1)

          0 Existing Customer    New Customer
  0.1415441      0.3455882       0.5128676
```

- *Takeaway 1*: New customers are giving the highest MRR and form the largest chunk of the business (51 percent).

```
> ggplot(market2, aes(x=Quarter, y=MRR.thou)) + geom_bar(stat="identity")
```

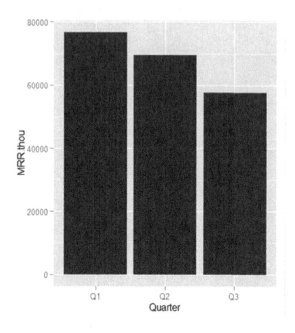

```
> table1<- table(market2$CustomerType, market2$Quarter)

> table1

                   Q1  Q2  Q3
  0                  0  77   0
  Existing Customer 56  98  34
  New Customer      143  98  38

> prop.table(table1,2)

                         Q1        Q2        Q3
  0                 0.0000000 0.2820513 0.0000000
  Existing Customer 0.2814070 0.3589744 0.4722222
  New Customer      0.7185930 0.3589744 0.5277778
```

- *Takeaway 2*: The quarter-wise MRR is showing a decreasing trend. The count of new customers is highest in Q1, and this is driving the trend.

Check the average MRR across customer type. You are looking at quarter-wise trends summarized over customers.

```
> ggplot(market2, aes(x=factor(CustomerType), y=MRR.thou)) + stat_summary(fun.y="mean",
geom="bar")
```

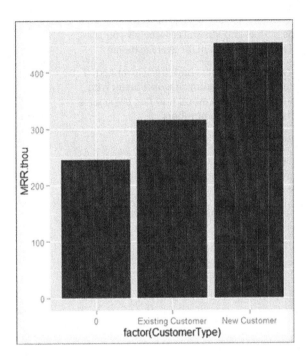

- *Takeaway 3*: You can see that the average MRR for new customers is the highest.

- *Assignment*: Do a similar exploration for verticals (through the column newvar) and create some takeaway points.

Using Descriptive Statistics

Let's look at how you can use some simple statistics to understand the data better. You have already come across some of these statistics in the earlier case study. Now you will be formally introduced to them.

Measures of Central Tendency

Let's look at measures of central tendency:

- *Mean*: The mean is the average score of the sample or set of numbers. Add up all the values and then divide the number by the company. Mean helps you understand the average normal value for a set of numbers. Thus, you can use it to describe the average age, the average income, the average score in the test, and so on.

It is to be noted that sometimes the mean cannot be trusted to give the complete picture. For example, the mean of 0 and 10 is 5, and the mean of 4 and 6 is also 5. You need some more information to get a good understanding of the data. This is covered later in this chapter.

What can you do with an average/mean?

You can then take an individual value and compare it to the average to draw some conclusions. If the average height of Indians is 5 feet and Mr. Anil is 6 feet tall, you can conclude that, in comparison, Mr. Anil is much taller than the average Indian.

- *Median*: This is the middle score/number after the numbers have been arranged in numerical order. The order could be ascending or descending. It works better with odd numbers since with odd numbers you will have one central number. If you have even numbers, you can use the average of the middle two numbers.

How can you use the median or central number?

Median provides you with an understanding of the normal/central value for a series. The advantage of median is that it reduces the effect of extreme values (also called *outliers*).

How are mean and median different? When should you use mean, and when should you use median?

Look at three numbers: 4, 5, 6.

```
mean = (4+5+6)/3 = 5
median = 5
```

Now look at three other numbers: 0, 1, 1, 1, 10.

```
mean = (0+1+1+1+20)/5 = 4.6
median = 1
```

Thus, mean is influenced by extreme values, also called *outliers*. The median then gives a more accurate picture of the middle, central, or most likely value for this series of numbers.

- *Mode*: This is defined as the most frequently occurring number in a series of number. It is especially useful when you are dealing with variables that are divided into a fixed number of categories. This is generally between three to five categories. An example is customer service feedback that can have any value from 1 to 5 or an employee rating that can have any value from 1 to 4.

In conclusion, mean, median, and mode are the three common measures of central tendency. It is always easier and more accurate to make conclusions about the most likely value for a sample data set when you see the mean and median together. Use the mode when the variable has three to five categories only.

Statistical dispersion tells you how spread out all the measurements or observations for a particular variable are in a data set.

- *Range*: This is the measurement of the smallest value to the largest one, that is, minimum to maximum.

Where can you use range?

You can use it to understand the boundaries of a data set/variable.

- *Interquartile range*: This is a measure of the boundaries of the four quartiles. How can you divide data into folk quartiles? You can sort the variable in sequence. Then pick the minimum to the 25th percentile value to define the quartile Q1; the 25th to 50th percentile to define the quartile Q2 (this is also known as the median); the 50th to 75th percentile to define the quartile Q3; the 75th percentile to the maximum value to define the quartile Q4. It is to be noted that the quartile definition is on the place that value holds. Thus, the concept of the median is followed while defining the quartiles.

Where can you use quartiles?

Quartiles are a useful measure of spread as they are much less affected by outliers or skewness (the concentration of data in one particular part of a density plot). Thus, they are often reported along with the median, especially when dealing with skewed data or data with outliers.

What Is Variation in Statistics?

Variation in data is measured using variance and standard deviation.

Absolute and Mean Absolute Deviation

These show the amount of deviation or radiation that occurs around the mean/average value. To find the variability in the observations, simply add up the deviation (distance from the mean) for each observation. The average deviation can then be calculated by dividing the sum of all deviations that might account for observations.

Calculating the absolute deviation is simple. To find out the total variability, you can sum the observation value minus the mean value for all the observations. However, you will get both positive and minus signs, and when you add all of this, you will get a total deviation of zero. As you are interested only in the deviation of the score and not really interested with the values above or below the mean value, you can ignore the negative numbers and only look at the absolute numbers, giving you the absolute deviation. Adding up all of these absolute deviations and dividing them by the total number of observations will give you the mean absolute deviation.

$$\text{Mean Absolute Deviation (MAD)} = \frac{\sum_{i=1}^{n} |X_i - \mu|}{n}$$

```
X = observation value
μ= Mean
N = count of observations
|| = absolute value
Σ = sum / addition
```

Say you have a set of five employees with the ages 20, 30, 35, 40, and 45, as shown in Figure 4-5.

Absolute deviation

Age of employee	Deviation (Obs - Mean)	Absolute Deviation
20	-14	14
30	-4	4
35	1	1
40	6	6
45	11	11

Mean	34
Sum of Deviation	0
Sum of Absolute Deviation	36
Mean Absolute Deviation	7.2

Figure 4-5. *Absolute deviation and mean absolute deviation*

Variance

This is a more common measure of deviation or the spread of data around the mean.

The deviation of each point from the mean is calculated. Unlike in absolute deviation, where you used the absolute value of the deviation to read the negative values, in variance you achieve the positive values by squaring each of the deviations. Add the squared deviations to get the sum of squares. Then divide this number by the total count of observations in your group of data. See Figure 4-6.

$$\sigma^2 = \frac{\sum (X - \mu)^2}{N}$$

```
X = observation value
μ= Mean
N = count of observations
Σ = sum / addition
```

Variance

Age of employee	Deviation (Obs - Mean)	Variance (Deviation ^2)
20	-14	196
30	-4	16
35	1	1
40	6	36
45	11	121

Mean	34
Total Variance	370
Variance	74

Figure 4-6. *Variance is the total variance divided by the count of observations*

Interpreting variance is easy. If the observations are clustered around the mean, the variance will be small. Conversely, if the observations are spread out, the variance will be large. However, there are some difficulties while interpreting variance. See Figure 4-7.

Mean	34
Total Variance	370
Variance	74
Standard Deviation	8.602325267

Figure 4-7. *Standard deviation*

- Because the deviations are squared, if the data contains outliers, one or two outliers can significantly increase the variance.

- The variance is measured in the units squared. This makes it difficult to be measured in any unit.

- To overcome these two problems, calculating the standard deviation is a good way out.

- The standard deviation is a measure of how spread out the numbers are.

- Its symbol is σ (the Greek letter sigma).

- The formula is easy: it is the square root of the variance.

The advantage of standard deviation is that it is measured in the same quantity as the unit of the variable. Thus, it is easy to use and easy to interpret.

The distribution of a statistical data set or a variable is a listing or function showing all the possible values or intervals of the data and how often they occur. Thus, I am talking about frequency distributions. When a distribution is organized, it is often ordered from the smallest to largest value. It is broken into reasonably sized groups and put into graphs and charts to examine the shape center and amount of variability in the data.

Thus, a histogram lends itself beautifully to understanding and working with the graphical output of a distribution, especially when the data is broken into categories. This is also called *discrete uniform* data.

If you want to see the distribution of a continuous set of data, a density plot enables you to do so. This is also called *continuous uniform* data.

When you start off with data that needs to be put into a distribution, you can answer the following four questions to help you:

- Is the data discrete, or is the data continuous?

- If the data symmetrical? If it is asymmetrical, which direction does it lie in?

- Does the data have upper or lower limits?

- What is the likelihood of observing extreme values in the data?

Here are some ways to define distribution:

- *Binomial distribution*: This measures the probability of the number of successes over a given number of events that are a specific probability of success in each try. An example is tossing a coin ten times where there are only two possible outcomes: head/tails. In business, an example is the attrition of employees where there are only two possible outcomes: attrition/no attrition.

- *Poisson distribution*: This measures the likelihood of a number of events occurring within a given time interval. Here the key parameter that is required is the average number of events in a given interval. An example is incoming phone calls in a service center.

- *Geometric distribution*: This measures the likelihood of when the first success will occur. Thus, when I toss a coin, there's a 50 percent chance that I will get a head in the first try and then a 12.5 percent chance that I will get a head in the third try too. When this is plotted, it leads to a geometric distribution.

- *Hypergeometric distribution*: This measures the probability of a specific number of successes in N trials without replacement for a finite population.

- *Discrete uniform distribution*: This is the simplest distribution and applies when all outcomes have the equal probability of occurring.

- *Normal distribution*: This is also called the bell curve. It refers to a family of continuous probability distributions. If a variable is a continuous variable, its probability distribution will be a continuous probability distribution.

CHAPTER 5

■ ■ ■

Visualization

Do you get overwhelmed looking at rows and rows of data? If yes, then this is a chapter that will help you make sense of the data by representing it as visuals, in other words, graphs and charts.

What Is Visualization?

Visualization is a process or technique of creating images or diagrams to communicate messages. These can include infographics and animations.

Prehistoric man made images on stone and clay tablets to create maps and show directions and for better navigation. Trade routes were through the seas and across desserts and through jungles. To improve governance and collection of revenues, the kings and traders of Egypt, Rome, and Greece had extensive maps made from 1500 to 300 BC. Thus, information visualization is an old art.

William Playfair (1759–1823) was a Scottish engineer and political economist. He was the founder of the graphical methods of statistics. He made the first modern representation of data using a pie chart, bar chart, and line graph (see Figure 5-1).

© Subhashini Sharma Tripathi 2016

S. S. Tripathi, *Learn Business Analytics in Six Steps Using SAS and R*, DOI 10.1007/978-1-4842-1001-7_5

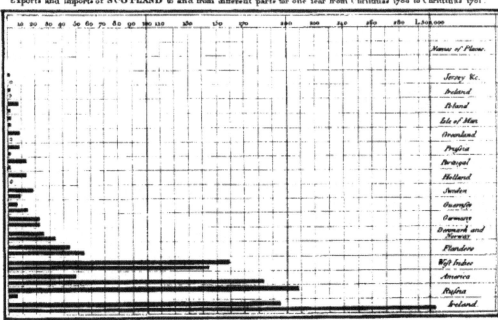

Figure 5-1. An early visualization (source: http://visually.visually.netdna-cdn.com/ExportsandImpor tsofScotland_4f1454ffef0be_w587.gif)

With the line chart, Playfair showed that it is not just the lines themselves that convey meaning (refer to Figure 5-2).

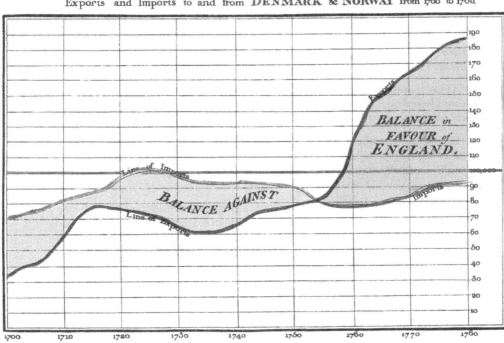

Figure 5-2. An early line graph (source: http://visually.visually.netdna-cdn.com/ExportsandImportst oandfromDenmarkNorwayfrom1700to1780_4e4d6b78e9bb9_w587.png)

■ **Note**   Are you interested in seeing the greatest visualizations in history? Look at "12 Great Visualizations That Made History" by Drew Skau at http://blog.visual.ly/12-great-visualizations-that-made-history/. Also look up "Milestones in the History of Data Visualization" from the original *Milestones in the History of Thematic Cartography, Statistical Graphics, and Data Visualization*, an illustrated chronology of innovations by Michael Friendly and Daniel J. Denis (York University, Canada). You can download it from www.math.yorku.ca/SCS/Gallery/milestone/Visualization_Milestones.pdf.

Data Visualization in Today's World

Information visualization as we understand it today became mainstream in the 1900s. The following are important points:

- Data visualization was used first in textbooks, and the study of graphs and charts became standard parts of science books and study curriculums.

- A French cartographer named Jacques Bertin published his work in *Semiologie Graphique*. This is considered the theoretical foundation of the stream of data visualization.

- The next stage was the advent and use of technology and computers. This era saw the creation of specialized visualization software and programs. Also, visualizations, graphs especially, started being seen in newspapers and magazines, and MIS and BI became subjects in MBA/PGDBA curriculum.

- 2002 saw the advent of the cloud revolution. Better and more advanced tools have helped create huge visualization engines that run over large data repositories called *enterprise-wide datawarehouses* (EDWs). These huge repositories started feeding the business intelligence interphases (QlickView, Cognos, SAS BI Studio, and so on) and churning out sharper, interactive visuals. However, it is true that Excel has emerged as the most cost-effective and user-friendly tool.

- Huge reports ensure that people get overwhelmed. Edward Tufte created the sparkline in 2004: a small, word-sized graphic that can be embedded into sentences, tables, headlines, spreadsheets, or graphics.

- The decade starting in 2011 has seen the emergence of interactive data visualization. Interactive visualization moves beyond the display of static graphics and spreadsheets to using computers and mobile devices to drill down into charts and graphs for more details and interactively (and immediately) changing what data you see and how it is processed. Thus, as the data gets added, so the graphs and visuals change.

Why Do Data Visualization?

Visualization is the step before hard-core statistical techniques are applied on the data. It serves two big purposes.

- Making sense of data

- Communicating (a picture is worth a thousand words)

Thus, it makes data and its interpretation easy on the eyes and the brain. View the following:

- *Figure 5-3*: The data

- *Figure 5-4*: The graph

Month of Year	Sales Amount	Total Product C...	Gross Profit Ma...	Gross Profit
January	1309863.2511	1046855.0401	0.20079058694...	263008.211
February	2451605.6244	2161789.71439...	0.11821473532...	289815.910000...
March	2099415.6158	1781531.84109...	0.15141536164...	317883.774700...
April	1546592.2292	1250946.0643	0.19115973772...	295646.164900...
May	2942672.90960...	2589467.20809...	0.12206783170...	359205.701500...
June	1678567.4193	2010739.60289...	-0.19789029012...	-332172.193599...
July	962716.741700...	754715.7636	0.21605625942...	208000.978100...
August	2044600.0034	1771778.75389...	0.13343502349...	272821.249500..
September	1639840.109	1393936.67389...	0.14995573882...	245903.43510001
October	1358050.4703	1124337.2647	0.17209463912...	233713.205600...
November	2868129.20330...	2561131.77409...	0.10703751729...	306997.42920002
December	2458472.4342	2085375.78659...	0.15175954076...	373096.647600...

Figure 5-3. *The data for this section*

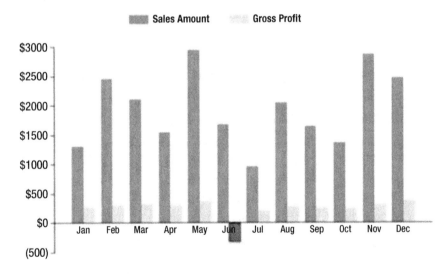

Figure 5-4. *The graph for Figure 5-3*

Question: Which of the two can you interpret easily?

I am sure you too will prefer Figure 5-4 (the graphical output). At one glance you can see that the gross profit was negative in the month of June and that the maximum revenue or sale amount was for the month of May. Thus, visuals help cognition or understanding, and therefore they help in decision-making for businesses.

The Gestalt theory of visual perception refers to theories of visual perception developed by German psychologists in the 1920s. (*Gestalt* is a psychology term that means "unified whole.") These theories attempt to describe how people tend to organize visual elements into groups or unified wholes.

Basically, our minds can put together small parts of a diagram to create a unified whole. A simple well-defined object is recognized much more easily than a detailed object with a heart to recognize contour.

Our mind also fills in the gaps. So, lines are as good as full lines.

All these principles are used by visualization software to create various types of graphs and charts that managers can call on when they're creating the visuals and presentations.

What Are the Common Types of Graphs and Charts?

So, what are the common types of graphs and charts? And what does each one of them depict?

- *Composition*: If you want to depict a whole, or 100 percent, and the segments that are within it, you can use the following:

    - A pie chart, which will show a simple share of the total.

    - A waterfall chart, which will show the accumulation of small parts to create the total.

    - A stacked 100 percent column chart with some complements. Each stack will represent 100 percent, and each component will show up in different colors. You can use this to compare a share of the total across multiple time frames or segments.

- *Distribution*: If you want to show the distribution/count per segment of data, you can use the following charts:

    - A histogram to plot frequency across categorical variables

    - A density plot to plot a concentration of variables that are continuous in nature

    - A scatter chart to plot a concentration of one variable against another (bivariate)

    - A 3D area chart to depict three variables against each other

- *Comparison*: When you want to compare data across segments or time periods, you can use the following graphs:

    - A line graph to depict the movement of various variables/categories across periods of time

    - A column chart to depict a few categories and compare data

- *Relationships*: When you want to explore the relationship between variables in you data set, you can use the following graphs:

    - A scatter plot to see the relationship between two variables (bivariate)

    - A bubble chart to see the relationship between three variables

Case Study on Graphs and Charts Using SAS

Fraud is the misappropriation of funds. This case study deals with understanding the data through graphs and charts to pinpoint possible fraud cases.

About the Data

The data consists of a council's contracts register. This lists contract expenditures that are more than $5,000.

What Is This Data?

This data set lists details of all procurement spending greater than $5,000, including purchase orders and tenders.

Definitions

The data set contains the following fields that you should use:

Effective Date	Date of data pull
Contract Reference Number	Contract id
Service responsible	Area of contract
Contract Start Date	Start date
End Date	End date (if missing, assume ongoing)
Review Date	Date of supervisory review
Extension Period	No of days
Contract Value	Amount

Problem Statement

Do an exploratory data analysis to understand the data better. Describe the date and number fields and understand the distributions.

- Create contingency tables, graphs, and charts to better understand the data.
- Do some projects finish significantly earlier than the date expected?
- What is the per-day rate on average for projects?
- Segregate the cases that should be put under investigation for fraud.

Fraud in these contracts can be of two types.

- The project can be completed in, say, 10 days but the contractor quotes 25 days. He then finishes the work in 10 days and pockets the money for the other 15 days.
- The project can be completed at $1 per day of work, and the contractor charges $5 per day of work .

Solution in SAS

Let's see how to solve this problem using the SAS tool.

First you will divide the tasks as per the DCOVA and I methodology.

1. Define. The project is to create graphs and charts to understand cases that could signify fraud defined as follows:

    *   Lower than normal time to complete the project

    *   Higher than normal cost per delivery day of project

2. Collect. The data is given. No other data is required to be referenced.

3. Organize.

    *   Missing values

    *   Outliers

    *   Non-numeric to numeric data

4. Visualize the y variables across significant segments of Service Responsible.

SAS Code and Solution

You will now execute the plan of DCOVA and I (as scripted earlier) through SAS code.

Import data. To start working on the data, you need to bring the data into the SAS system.

Please note that /* */ are the symbols used to mark a comment, which is a description of the code.

```
/* IMPORT DATA USING INFILE INFORMAT STATMENT */
/* THIS CODE WILL GET AUTO GENERATED IF YOU USE THE IMPORT DATA BUTTON
UNDER THE FILE TAB ON TOP LEFT CORNER OF SCREEN*/
DATA WORK.CONTRACTS;
    LENGTH
        Effective_Date    $ 11
        Contract_Reference_Number $ 8
        Title_of_agreement $ 94
        Service_responsible $ 49
        Description_of_goods_and_service $ 94
        Contract_Start_Date    8
        End_Date          $ 10
        Review_Date       $ 10
        Extension_Period   8
        Contract_Value     8
        SupplierName      $ 40
        NominatedContactPoint $ 37
        ;
    LABEL
        Effective_Date    = "Effective Date"
        Contract_Reference_Number = "Contract Reference Number"
        Title_of_agreement = "Title of agreement"
        Service_responsible = "Service responsible"
        Description_of_goods_and_service = "Description of goods and services"
        Contract_Start_Date = "Contract Start Date"
```

```
      End_Date            = "End Date"
      Review_Date         = "Review Date"
      Extension_Period = "Extension Period"
      Contract_Value      = "Contract Value" ;
FORMAT
      Effective_Date     $CHAR11.
      Contract_Reference_Number $CHAR8.
      Title_of_agreement $CHAR94.
      Service_responsible $CHAR49.
      Description_of_goods_and_service $CHAR94.
      Contract_Start_Date MMDDYY10.
      End_Date            $CHAR10.
      Review_Date         $CHAR10.
      Extension_Period BEST2.
      Contract_Value      BEST11.
      SupplierName        $CHAR40.
      NominatedContactPoint $CHAR37.
   ;
INFORMAT
      Effective_Date     $CHAR11.
      Contract_Reference_Number $CHAR8.
      Title_of_agreement $CHAR94.
      Service_responsible $CHAR49.
      Description_of_goods_and_service $CHAR94.
      Contract_Start_Date MMDDYY10.
      End_Date            $CHAR10.
      Review_Date         $CHAR10.
      Extension_Period BEST2.
      Contract_Value      BEST11.
      SupplierName        $CHAR40.
      NominatedContactPoint $CHAR37.
      ;
INFILE '/saswork/SAS_work6710000127DB_odaws02-prod-sg/#LN00030'
      LRECL=332
      ENCODING="UTF-8"
      TERMSTR=CRLF
      DLM='7F'x
      MISSOVER
      DSD ;
INPUT
      Effective_Date     : $CHAR11.
      Contract_Reference_Number : $CHAR8.
      Title_of_agreement : $CHAR94.
      Service_responsible : $CHAR49.
      Description_of_goods_and_service : $CHAR94.
      Contract_Start_Date : ?? MMDDYY10.
      End_Date            : $CHAR10.
      Review_Date         : $CHAR10.
      Extension_Period : ?? BEST2.
      Contract_Value      : ?? COMMA11.
      SupplierName        : $CHAR40.
```

```
          NominatedContactPoint : $CHAR37.
          ;
RUN;

/* EXPLANATION:-
LENGTH - sets the length for each variable ; $ symbol shows it's a character variable
LABEL - sets the explanation for the variables; can be seen in the PROC CONTENTS output
FORMAT - controls written appearance of the variable values
INFORMAT - reads data into SAS
INFILE - gives details about the datafile that we are importing (external file)
DSD (delimiter-sensitive data)
LRECL=logical-record-length specifies the logical record length.
MISSOVER =prevents an INPUT statement from reading a new input data record if it does not
find values in the current input line for all the variables in the statement.
ENCODING= specifies the encoding to use when reading from the external file (this is
optional )
INPUT - creates input record in input buffer (from where it will be inserted in a sas
datafile)*/

PROC PRINT DATA=WORK.CONTRACTS (OBS=10); RUN;
NOTE ALTERNATE IMPORT CODE :-
FILENAME REFFILE "/home/subhashini1/my_content/Contacts register Aug 2015 and purchase order
over 5000 April to June 2015.csv" TERMSTR=CR;

PROC IMPORT DATAFILE=REFFILE
    DBMS=CSV
    OUT=WORK.IMPORT
       REPLACE;
    GETNAMES=YES;
RUN;
DATA WORK.CONTRACTS;
SET WORK.IMPORT;RUN;
```

View the contents of the SAS file and understand the variables. Most of the time, the data sets will be very large and it may not be possible to open them to view in Excel or another tool. At this stage you will get a sense of the rows and columns in the data.

```
PROC CONTENTS DATA=WORK.CONTRACTS; RUN;
```

	Alphabetic List of Variables and Attributes					
#	Variable	Type	Len	Format	Informat	Label
2	Contract_Reference	Char	8	$CHAR8.	$CHAR8.	Contract Reference Number
6	Contract_Start_Date	Num	8	MMDDYY	MMDDYY10.	Contract Start Date
10	Contract_Value	Num	8	BEST11.	BEST11.	Contract Value
5	Description_of_good	Char	94	$CHAR94.	$CHAR94.	Description of goods and services
1	Effective_Date	Char	11	$CHAR11.	$CHAR11.	Effective Date
7	End_Date	Char	10	$CHAR10.	$CHAR10.	End Date
9	Extension_Period	Num	8	BEST2.	BEST2.	Extension Period
12	NominatedContactP	Char	37	$CHAR37.	$CHAR37.	
8	Review_Date	Char	10	$CHAR10.	$CHAR10.	Review Date
4	Service_responsible	Char	49	$CHAR49.	$CHAR49.	Service responsible
11	SupplierName	Char	40	$CHAR40.	$CHAR40.	
3	Title_of_agreement	Char	94	$CHAR94.	$CHAR94.	Title of agreement

■ **Note**   You can easily copy and paste the output from the Results window into an Excel sheet. You can then use the output to create graphs and charts in Excel if you want.

Keep the required variables. When you do any project, you have a definitive goal that has been clearly spelled out in the define part of the DCOVA and I process. You therefore want to look at fields and variables that have a bearing on the problem defined.

```
/* KEEP ONLY THOSE VARIABLES THAT ARE AVAILABLE IN THE
DEFINITIONS GIVEN IN THE PROJECT*/

DATA WORK.CONTRACTS1 (KEEP=Contract_Reference_Number
Contract_Start_Date
Contract_Value
End_Date
Extension_Period
Review_Date
Service_responsible);
SET WORK.CONTRACTS; RUN ;

PROC PRINT DATA=WORK.CONTRACTS1 (FIRSTOBS=70 OBS=75); RUN ;
```

Create the required fraud indicator variables (y). At this stage you want to create the y variable since the definition has been understood in the define stage.

```
/* CREATE Y VARIABLES
TAT = END DATE - START DATE
PERDAY = CONTACT AMOUNT / TAT

NOTE - END DATE IS CHARACTER (AS SEEN IN PROC CONTENTS)*/
```

```
DATA WORK.CONTRACTS1 ;
SET WORK.CONTRACTS1 ;
END_DATE1=INPUT(End_Date, MMDDYY10.);
FORMAT END_DATE1 MMDDYY10.; RUN;

PROC PRINT DATA=WORK.CONTRACTS1 (OBS=10); RUN;

PROC CONTENTS DATA=WORK.CONTRACTS1; RUN;
```

You should look at the output to understand whether the modification you wanted and coded for has run. Thus, you can study the output in the following image:

Alphabetic List of Variables and Attributes						
#	Variable	Type	Len	Format	Informat	Label
1	Contract_Reference	Char	8	$CHAR8.	$CHAR8.	Contract Reference Number
3	Contract_Start_Date	Num	8	MMDDYY	MMDDYY10.	Contract Start Date
7	Contract_Value	Num	8	BEST11.	BEST11.	Contract Value
8	END_DATE1	Num	8	MMDDYY10.		
4	End_Date	Char	10	$CHAR10.	$CHAR10.	End Date
6	Extension_Period	Num	8	BEST2.	BEST2.	Extension Period
5	Review_Date	Char	10	$CHAR10.	$CHAR10.	Review Date
2	Service_responsible	Char	49	$CHAR49.	$CHAR49.	Service responsible

Here you need to create a new field called TAT, which is a derived field.

```
DATA WORK.CONTRACTS1;
SET WORK.CONTRACTS1;
TAT= END_DATE1-Contract_Start_Date;
PERDAY = Contract_Value/TAT; RUN;

PROC PRINT DATA=WORK.CONTRACTS1 (OBS=10);
VAR TAT PERDAY; RUN;
```

Collect and organize. This is the stage of collecting the data (if this is required) and organizing it.

```
/* C - NO ACTIVITY REQUIRED
O - MISSING VALUES
CHECK MISSING VALUES FOR NUMERIC VARIABLES USING PROC MEANS
CHECK MISSING VALUES FOR CHARACTER VARIABLES USING PROC FREQ*/

PROC MEANS DATA=WORK.CONTRACTS1; RUN;
```

The MEANS Procedure

Variable	Label	N	Mean	Std Dev	Minimum	Maximum
Contract_Start_Date	Contract Start Date	194	19968.74	457.330218	17623	20362
Extension_Period	Extension Period	84	0.75	1.7276978	0	14
Contract_Value	Contract Value	194	702287.9	4524217.01	0	49000000
END_DATE1		83	20936.33	671.611914	20331	25383
TAT		83	1290.34	716.242691	363	5478
PERDAY		83	1167.87	6794.94	0	59157.89

You want to look at the frequency distribution for a variable by using the SAS code.

```
PROC FREQ DATA=WORK.CONTRACTS1;
TABLES Service_responsible; RUN;
```

The FREQ Procedure

Service responsible

Service_responsible	Frequency	Percent	Cumulative Frequency	Cumulative Percent
CHIEF EXECUTIVE	1	0.52	1	0.52
COMMUNITY	21	10.82	22	11.34
Community	3	1.55	25	12.89
DEMOCRATIC LEGAL AND POLICY	12	6.19	37	19.07
Democratic Legal & Policy	2	1.03	39	20.1
ENVIRONMENT	10	5.15	49	25.26
Environment	12	6.19	61	31.44
FINANCE & COMMERCIAL	23	11.86	84	43.3
Finance & Commercial	31	15.98	115	59.28
HOUSING	5	2.58	120	61.86
HUMAN RESOURCES ICT & SHARED SUPPORT	19	9.79	139	71.65
Human Resources ICT/CSC & Shared Support Services	25	12.89	164	84.54
PLANNING	8	4.12	172	88.66
PROPERTY	11	5.67	183	94.33
Planning & Sustainability	11	5.67	194	100

You want to remove variables that won't be used in the analysis.

```
DATA WORK.CONTRACTS1 (DROP= End_Date
Extension_Period
```

```
Review_Date);
SET WORK.CONTRACTS1; RUN;
/* IN THE PROC FREQ OUTPUT WE CAN SEE THAT BECAUSE OF CASE DIFFERENCE
SOME SEGMENTS ARE COMING TWICE EG :- COMMUNITY / Community;
CONVERT ALL THE DATA TO UPPER CASE TO STANDARDISE OUTPUT

WE CAN ALSO SEE THAT SOME MINOR SPELLING DIFFERENCE IN CAUSING DIFFERENT
SEGMENTS EG:- DEMOCRATIC LEGAL AND POLICY /Democratic Legal & Policy.
CREATE SAME SPELLING ACROSS */

DATA WORK.CONTRACTS1;
SET WORK.CONTRACTS1;
IF Service_responsible IN ('Democratic Legal & Policy')
THEN Service_responsible= 'DEMOCRATIC LEGAL AND POLICY';
IF Service_responsible IN ('Human Resources ICT/CSC & Shared Support Services')
THEN Service_responsible= 'HUMAN RESOURCES ICT & SHARED SUPPORT';
IF Service_responsible IN ('Planning & Sustainability')
THEN Service_responsible= 'PLANNING';
RUN;

DATA WORK.CONTRACTS1;
SET WORK.CONTRACTS1;
Service_responsible= UPCASE(Service_responsible); RUN;

PROC FREQ DATA=WORK.CONTRACTS1;
TABLES Service_responsible; RUN;
```

The table output shown next shows the breakdown of the number of cases for each type/segment of Service Responsible. It allows you to understand which are the largest segments of Service Responsible.

The FREQ Procedure

Service responsible

Service_responsible	Frequency	Percent	Cumulative Frequency	Cumulative Percent
CHIEF EXECUTIVE	1	0.52	1	0.52
COMMUNITY	24	12.37	25	12.89
DEMOCRATIC LEGAL AND POLICY	14	7.22	39	20.1
ENVIRONMENT	22	11.34	61	31.44
FINANCE & COMMERCIAL	54	27.84	115	59.28
HOUSING	5	2.58	120	61.86
HUMAN RESOURCES ICT & SHARED SUPPORT	44	22.68	164	84.54
PLANNING	19	9.79	183	94.33
PROPERTY	11	5.67	194	100

Visualization

You'll start with a bar chart to understand the average value of TAT across different segments of Service Responsible.

The first method is to use the buttons at the top of the screen (the second method is writing SAS code).

The sequence of the button-driven graph creation is Tasks ➤ Graph ➤ Bar Chart Wizard.

This will open the wizard. Just use the drag-and-drop functionality to create the graph.

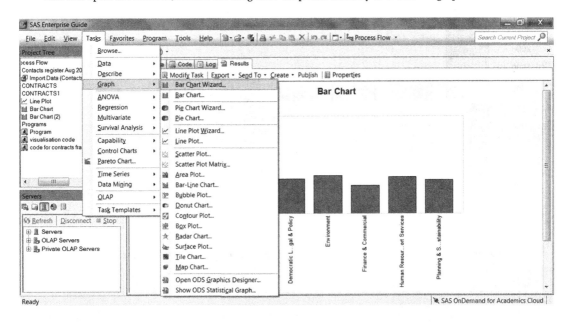

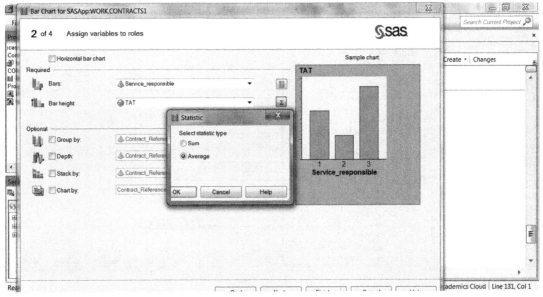

The bar chart appears.

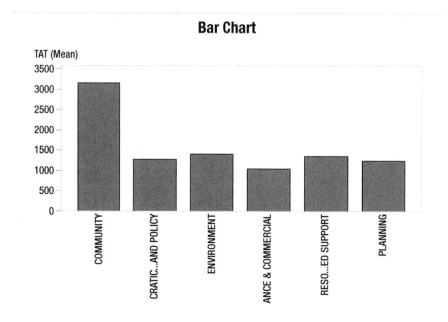

Let's see how to create a pie chart in SAS so that you can see the percentage of contribution of each segment.

The sequence of using the buttons is Tasks ➤ Graph ➤ Pie Chart Wizard.

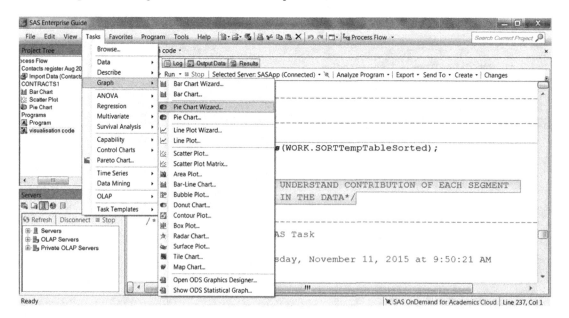

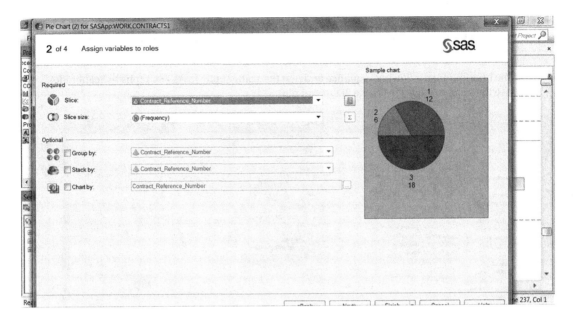

The code gets auto-generated in the code window. This is a good feature, and you can copy and paste the code to keep it for future use.

The output looks like the following:

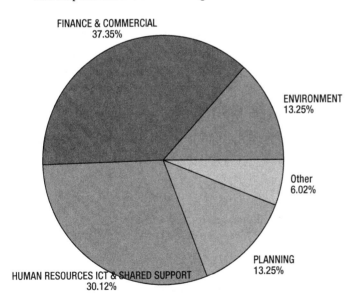

You can use the scatter plot to check the distribution of TAT and also to see whether there are any significantly higher/lower values (called *outliers*) in the data.

■ **Note** I am doing this for the variable TAT. You should do it for the variable Per Day Rate as an exercise.

Use the buttons in the following sequence and plot the scatter plot: Tasks ➤ Graphs ➤ Scatter Plot. Choose the same variable (TAT) for a 2D scatter plot.

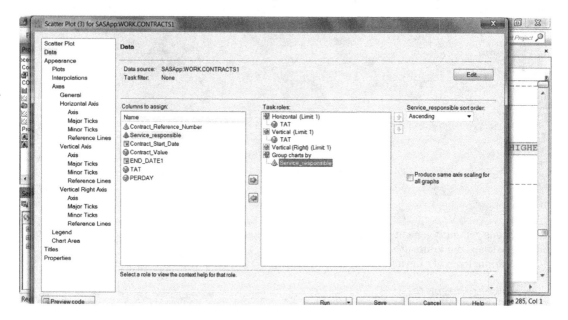

Export the result as a PDF. This is a user-friendly feature in SAS.

Case Study on Graphs and Charts Using R

Fraud is the misappropriation of funds. This case study deals with understanding the data through graphs and charts to pinpoint possible fraud cases.

About the Data

This is a council's contracts register; it lists contract expenditures greater than $5,000.

What Is This Data?

This data set lists the details of all procurement spending more than $5,000, including purchase orders and tenders.

Definitions

The data set contains the following fields:

Effective Date	Date of data pull
Contract Reference Number	Contract id
Service responsible	Area of contract
Contract Start Date	Start date
End Date	End date (if missing, assume ongoing)
Review Date	Date of supervisory review
Extension Period	No of days
Contract Value	Amount

Problem Statement

Do an exploratory data analysis to understand the data better. Describe the date and number fields and understand the distributions.

- Create contingency tables, graphs, and charts to better understand the data.
- Do some projects finish significantly earlier than the date expected?
- What is the per-day rate on average for projects?
- Segregate the cases that should be put under investigation for fraud.

Fraud in these contracts can be of two types.

- The project can be completed in, say, 10 days but the contractor quotes 25 days. He then finishes the work in 10 days and pockets the money for the 15 days.
- The project can be completed at $1 per day of work, and the contractor charges $5 per day of work.

Solution in R

Let's make a note of how you will approach this project using the DCOVA and I methodology.

- Define. The project is to create graphs and charts to understand cases that could signify fraud defined as follows:
    - Lower than normal time to complete the project
    - Higher than normal cost per delivery day of project
- Collect. The data is given. No other data is required to be referenced.
- Organize.
    - Missing values
    - Outliers
    - Non-numeric to numeric data
- Visualize the *y* variables across significant segments of Service Responsible.

R Code and Solution

You will now execute the process described earlier using code in R. You start the process by bringing or importing the data into R.

To import data, first set the location where the data exists.

```
setwd('H:/springer book/Case study/CaseStudy3')
```

```
#IMPORT DATA
mydata <- read.table("H:/springer book/Case study/CaseStudy3/Contacts register Aug 2015 and
purchase order over 5000 April to June 2015.csv", header=TRUE, sep=',')
```

Let's see how the data looks by using dim.

```
dim(mydata)
```

```
> dim(mydata)
[1] 194   17
```

To see how the data looks across rows and columns, use this:

```
View(mydata)
head(mydata, n=10)
```

An alternative method is to use the button Tools ➤ Import Dataset ➤ From Text File.

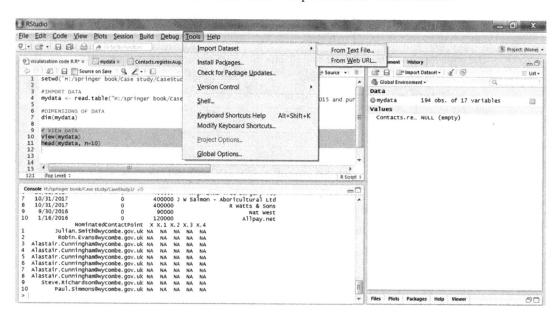

View the contents of the R file and look at the variables so that you can understand the formats in which each variable exists.

```
# VIEW CONTENTS
str(mydata)
> str(mydata)
'data.frame':     194 obs. of  17 variables:
 $ Effective.Date                : chr "August 2015" "August 2015" "August 2015" "August
                                    2015" ...
 $ Contract.Reference.Number     : chr "342" "228" "289" "290" ...
 $ Title.of.agreement            : chr "Air quality Equipment Maintenance" "ANPR for
                                    car parks" "Arboricultural Services" "Arboricultural
                                    Services" ...
 $ Service.responsible           : chr "Environment" "Environment" "Planning &
                                    Sustainability" "Planning & Sustainability" ...
 $ Description.of.goods.and.services: chr "Air quality Equipment Maintenance" "ANPR for
                                    car parks" "Arboricultural Services" "Arboricultural
                                    Services" ...
 $ Contract.Start.Date           : chr "1/1/2015" "2/26/2013" "11/1/2014" "11/1/2014" ...
 $ End.Date                      : chr "12/31/2015" "2/25/2018" "10/31/2018" "10/31/2018"
                                    ...
 $ Review.Date                   : chr "11/30/2015" "10/1/2017" "10/31/2017" "10/31/2017" ...
 $ Extension.Period              : int 0 2 0 0 0 0 0 0 0 0 ...
 $ Contract.Value                : num 5944 1200000 400000 400000 400000 ...
 $ SupplierName                  : chr "ESU1 ltd" "APT Control Ltd - Veri-Park" "Advance
                                    Tree Services Ltd" "Glendale Countryside Ltd" ...
 $ NominatedContactPoint         : chr "Julian.Smith@wycombe.gov.uk" "Robin.Evans@
                                    wycombe.gov.uk" "Alastair.Cunningham@wycombe.gov.uk"
                                    "Alastair.Cunningham@wycombe.gov.uk" ...
 $ X                             : logi  NA NA NA NA NA NA ...
 $ X.1                           : logi  NA NA NA NA NA NA ...
 $ X.2                           : logi  NA NA NA NA NA NA ...
 $ X.3                           : logi  NA NA NA NA NA NA ...
 $ X.4                           : logi  NA NA NA NA NA NA ...
```

If you want to just know the headers of the various fields, you can use the ls command.

```
ls(mydata)
> ls(mydata)
 [1] "Contract.Reference.Number"        "Contract.Start.Date"
 [3] "Contract.Value"                   "Description.of.goods.and.services"
 [5] "Effective.Date"                   "End.Date"
 [7] "Extension.Period"                 "NominatedContactPoint"
 [9] "Review.Date"                      "    Service.responsible"
[11] "SupplierName"                     "Title.of.agreement"
[13] "X"                                "X.1"
[15] "X.2"                              "X.3"
[17] "X.4"
```

Keep only the relevant variables and drop the variables that you will not use for further analysis.

```
# DROP VARIABLES  13, 14,15, 16,17
newdata1 <- mydata[c( -13, -14,-15, -16,-17)]
head(newdata1, n=10)
ls(newdata1)
```

```
newdata1$Effective.Date<-NULL
newdata1$Description.of.goods.and.services<-NULL
newdata1$Extension.Period<-NULL
newdata1$NominatedContactPoint<-NULL
newdata1$Review.Date<-NULL
newdata1$SupplierName<-NULL
newdata1$Title.of.agreement<-NULL
```

Let's understand how the data has changed by using the `dim` and `str` code.

```
dim(newdata1)
> dim(newdata1)
[1] 194    5
```

```
str(newdata1)
```

```
> str(newdata1)
'data.frame':                194 obs. of  5 variables:
 $ Contract.Reference.Number: chr "342" "228" "289" "290" ...
 $ Service.responsible      : chr "Environment" "Environment" "Planning & Sustainability"
                               "Planning & Sustainability" ...
 $ Contract.Start.Date      : chr "1/1/2015" "2/26/2013" "11/1/2014" "11/1/2014" ...
 $ End.Date                 : chr "12/31/2015" "2/25/2018" "10/31/2018" "10/31/2018" ...
 $ Contract.Value           : num 5944 1200000 400000 400000 400000 ...
```

Create required fraud indicator variables (y). This is the business problem defined in mathematical terms in the define stage.

```
# CREATE Y VARIABLES
# TAT = END DATE - START DATE
# PERDAY = CONTACT AMOUNT / TAT

# CONVERT DATE VARIABLES INTO DATE FORMAT
newdata1$Contract.Start.Date1<- as.Date(newdata1$Contract.Start.Date, "%m/%d/%Y")
newdata1$End.Date1<- as.Date(newdata1$End.Date,"%m/%d/%Y")
str(newdata1)
```

Now you drop redundant variables. This is to remove the variables that will not be used later in the analysis.

```
newdata1$Contract.Start.Date<-NULL
newdata1$End.Date<-NULL
str(newdata1)
```

You use the `str` code to understand how the data set has changed.

```
> str(newdata1)
'data.frame':                165 obs. of  5 variables:
 $ Contract.Reference.Number: chr "342" "228" "289" "290" ...
 $ Service.responsible      : chr "Environment" "Environment" "Planning & Sustainability"
                               "Planning & Sustainability" ...
```

118

```
$ Contract.Value          : num 5944 1200000 400000 400000 400000 ...
$ Contract.Start.Date1     : Date, format: "2015-01-01" "2013-02-26" "2014-11-01" ...
$ End.Date1               : Date, format: "2015-12-31" "2018-02-25" "2018-10-31" ...
```

Let's create new variables. These will be the derived variables that will be used in the analysis. You use the str code to understand the modification in the data set.

```
newdata1$tat<- newdata1$End.Date1-newdata1$Contract.Start.Date1
newdata1$tat1<- as.numeric(newdata1$tat)
newdata1$perday<- newdata1$Contract.Value/newdata1$tat1
```

```
str(newdata1)
> str(newdata1)
'data.frame':              194 obs. of  8 variables:
 $ Contract.Reference.Number: chr "342" "228" "289" "290" ...
 $ Service.responsible       : chr "Environment" "Environment" "Planning & Sustainability"
                               "Planning & Sustainability" ...
 $ Contract.Value           : num 5944 1200000 400000 400000 400000 ...
 $ Contract.Start.Date1      : Date, format: "2015-01-01" "2013-02-26" "2014-11-01" ...
 $ End.Date1                : Date, format: "2015-12-31" "2018-02-25" "2018-10-31" ...
 $ tat                       : Class 'difftime'  atomic [1:194] 364 1825 1460 1460 1460 ...
                               .. ..- attr(*, "units")= chr "days"
 $ tat1                      : num 364 1825 1460 1460 1460 ...
 $ perday                    : num 16.3 657.5 274 274 274 ...
```

Now you will do the Collect and Organize parts in the DCOVA and I process.

```
# C - NO ACTIVITY REQUIRED
# O - MISSING VALUES
# CHECK MISSING VALUES - in R the missing values are imported as NA values
```

You now check for missing values.

```
sapply(newdata1, function(x) sum(is.na(x)))
```

```
> sapply(newdata1, function(x) sum(is.na(x)))
Contract.Reference.Number        Service.responsible        Contract.Value
                        0                          0                     0
       Contract.Start.Date1                  End.Date1                   tat
                        0                        110                   110
                      tat1                     perday
                       110                        110
```

Drop observations where the end date is missing; keep observations where the data is complete. You decide to address the missing values.

```
newdata2<- newdata1[complete.cases(newdata1[,5:8]),]
```

Check the frequency to understand segments in the Service.responsible variable. You want to understand how the subsegments contribute to the data.

You will use the package plyr. Before starting, you will need to install the package.

```
# INSTALL PACKAGE

install.packages('plyr')
library('plyr')
```

You create a frequency table to understand the subgroups.

```
table1<- count(newdata2, 'Service.responsible')
```

table1 gets created, but to view the output, you will need to write code.

```
table1
```

```
> table1
                                Service.responsible freq
1                                         Community    3
2                        Democratic Legal & Policy    2
3                                       Environment   11
4                              Finance & Commercial   31
5 Human Resources ICT/CSC & Shared Support Services   25
6                          Planning & Sustainability   11
```

Note: R is not case sensitive for observations within variables (unlike SAS).

Visualization

To understand the average value of TAT across different segments of the variable Service_responsible, you want to create a chart.

To understand the average value of TAT across different segments of the variable Service_responsible, you will write the following code:

```
barplot(with(newdata2, tapply(tat, Service.responsible, mean) ))
```

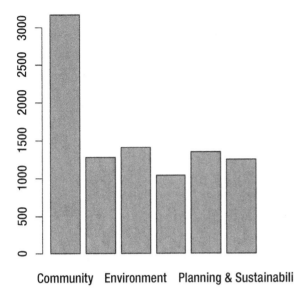

■ **Note** You should also explore the bar plot functions in the package ggplot2. You will have to start with installing the package.

Let's create a pie chart to understand the contribution of each segment

For Service_responsible in the data, you want to understand what percentage of the whole is contributed by each segment. You need to install the library MASS to continue since the package MASS has already been installed. If it is not installed for you, please start with the code INSTALL PACKAGES.

```
library(MASS)
```

Create the frequency table in the system.

```
serv.freq = table(newdata2$Service.responsible)
```

Convert the table output into a pie chart.

```
pie(serv.freq)
```

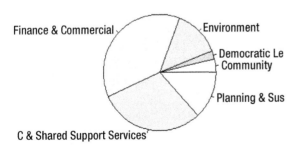

Draw a scatter plot to see whether some cases have significantly higher TAT or higher rate-per-day values. You want to identify where most of the values are concentrated and whether there are any extreme values, also called *outliers*.

■ **Note**   I am doing it for TAT. You should do it for Per Day Rate as an exercise.

```
plot(newdata2$tat)
```

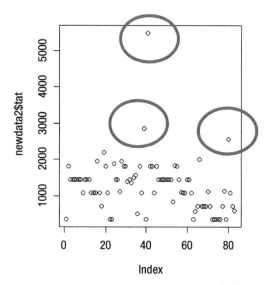

```
#scatter plot1
newdata3 <- subset(newdata2,newdata2$Service.responsible=='Community')
plot(newdata3$tat)
```

```
#scatter plot2
newdata4 <- subset(newdata2,newdata2$Service.responsible=='Democratic Legal & Policy')
plot(newdata4$tat)
```

```
#scatter plot3
newdata5 <- subset(newdata2,newdata2$Service.responsible=='Environment')
plot(newdata5$tat)
```

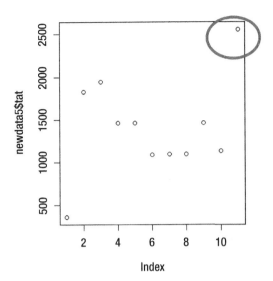

```
#scatter plot4
newdata6 <- subset(newdata2,newdata2$Service.responsible=='Finance & Commercial')
plot(newdata6$tat)
```

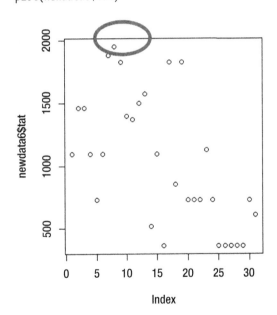

```
#scatter plot5
newdata7 <- subset(newdata2,newdata2$Service.responsible=='Human Resources ICT/CSC & Shared
Support Services')
plot(newdata7$tat)
```

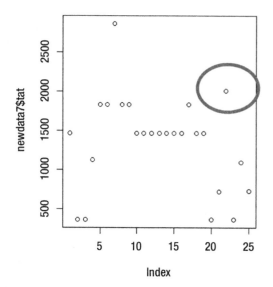

```
#scatter plot6
newdata8 <- subset(newdata2,newdata2$Service.responsible=='Planning & Sustainability')
plot(newdata8$tat)
```

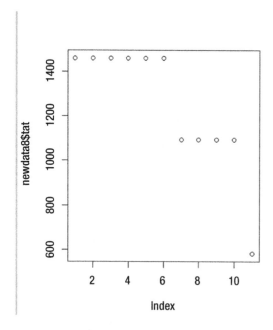

You have seen how to use visualizations to understand your data and answer some basic questions. Let's look at a common statistical technique that is used often to understand the associations within the data. Most of us use these terms in common, day-to-day language: correlation and covariance.

What Are Correlation and Covariance?

Correlation is the first measure of association between variables. What type of association am I talking about? I am talking about linear relationships. Thus, you will want to understand that if you increase or decrease X by a certain value, what is the effect on Y?

Will the Y variable increase when X increases and decrease when X decreases? If yes, then that is a positive correlation because the movement between the X variable and the Y variable is in the same direction. For example, as the summer heat increases, the number of mangoes in the market increases. Also, as the summer heat decreases with the onset of rains, the number of mangoes in the market decreases.

To summarize, correlation measures the relative strength of the linear relationship between two variables and has the following properties:

```
@ Unit-less
@ Ranges between -1 and 1
@ The closer to -1, the stronger the negative linear relationship
@ The closer to 1, the stronger the positive linear relationship
@ The closer to 0, the weaker any positive linear relationship
```

Covariance is the measure of how much two random variables change together. What is a random variable? Any field of variable that can take on a set of different values for different observations is called a random variable. Thus, all the fields in a data set that have multiple values are random variables.

What does changing together mean? If the greater values of one variable correspond with the greater values of the other variables or correspond with lower values of other variables, then they changed together. You can therefore conclude that the high values in one variable are reflected in the high values in the second variable or you can conclude that the high values in one variable are reflected in the lower values of second variable. In the first case, when the high values in one variable reflect the high values in the second variable, the covariance is positive. In the second case, when the high values in one variable reflect the low values in the second variable, the covariance is negative.

How is covariance different from correlation?

```
Co-relation is a scaled version of covariance.
cor(X,Y)=cov(X,Y) / sd(X)sd(Y)
(correlation of X,Y = Covariance of X,Y / (Standard Deviation X*Standard Deviation Y ))
```

Since correlation is derived from covariance, the two parameters always have the same sign (positive, negative, zero). Correlation is dimensional-less since the numerator and denominator have the same physical units. Covariance, on the other hand, is measured in the same units as the variables X and Y.

How to Interpret Correlation

The correlation is measured by the correlation coefficient. This correlation coefficient can be a number between 0 to 1. However, correlation has direction and so does the correlation coefficient. Hence, it is commonly said that correlation can be between -1 and +1. It is to be noted that -1 is not less than +1 when talking about correlation because the plus or minus reflects only the direction of the relationship.

It is interesting to note that the correlation coefficient will also exist for nonlinear/curvilinear relationships. A linear relationship has direct proportionality that causes variable 1 (Y) to change in equal values when variable 2 (X) changes. While in a nonlinear or curvilinear relationship, there is no proportionality between the dependent and independent variables that is not a consistent change in variable 1 (Y) when variable 2 (X) changes.

You will see a case study on correlation in the next chapter.

CHAPTER 6

Probability Using SAS and R

In this chapter, I will cover the concept of probability and how to calculate probability using the SAS and R tools. I will also cover the concept of distribution, especially normal distributions and how to work on distributions using SAS and R.

What Is Probability?

The chance of an event occurring is *probability*. What is the chance that it will rain today? What is the chance that you will reach the office in the next ten minutes? Given the existing grades, what is the chance that a student will pass the exam?

The *American Heritage Dictionary* defines the probability theory as the branch of mathematics that studies the likelihood of occurrence of random events in order to predict the behavior of defined systems. But many events can't be predicted with total certainty. The best you can say is how likely they are to happen, using the idea of probability.

> *Probability of an event happening = Number of ways it can happen / Total number of outcomes*

You probably use the word *probability* or *chance* in your day-to-day life. In statistics, the meaning of probability is the same. If, out of the last ten days, the bus has arrived late on eight days, what is the probability that the bus will be late today? It will be 8/10, which is an 80 percent probability that the bus will be late today.

The chances of rolling a 4 with a die can be figured out like this:

```
# Number of ways it can happen: 1 (there is only 1 face with a "4" on it)
# Total number of outcomes: 6 (there are 6 faces altogether)
```

So, the probability = 1/6.

What are the possible values of probability that an event will occur? The probability is always a number between 0 and 1 or between 0 percent and 100 percent; 0 means something cannot happen (impossible), and 1 (or 100 percent) means it is sure to happen.

What is the probability that an event will not occur? If you know the probability of an event occurring, it is easy to compute the probability that the event will not occur. If P(A) is the probability of Event A, then 1 − P(A) is the probability that the event does not occur. Thus, the probability of rolling a die to get a 6 as an outcome is one out of six possible outcomes.

© Subhashini Sharma Tripathi 2016
S. S. Tripathi, *Learn Business Analytics in Six Steps Using SAS and R*, DOI 10.1007/978-1-4842-1001-7_6

Probability of Independent Events: The Probability of Two or More Events

What is the probability of Event A *and* Event B occurring when only A or only B can occur at any one given time? For example, when you flip a coin, only heads or only tails can come up. You know the probability of getting heads when a coin is flipped is 1/2 (or .5). What is the probability of two heads to occur, in other words, first one head and then one head? The probability is the *Multiple* of the two individual probabilities: 1/2 * 1/2, or .5 * .5, which is 1/4 or .25.

If Events A and B are independent, then the probability of both A and B occurring is as follows:

```
P(A and B) = P(A) x P(B)
```

where P(A and B) is the probability of Events A and B both occurring, P(A) is the probability of Event A occurring, and P(B) is the probability of Event B occurring.

Now, what is the probability of Event A *or* Event B occurring? For example, if you flip a coin two times, what is the probability that you will get a head on the first flip or a head on the second flip (or both)? Letting Event A be a head on the first flip and Event B be a head on the second flip, then P(A) = 1/2, P(B) = 1/2, and P(A and B) = 1/4.

Therefore, P(A or B) = 1/2 + 1/2 - 1/4 = 3/4.

Probability of Conditional Events: The Probability of Two or More Events

In a pack of 52 cards, there are 4 aces. What is the probability of drawing any ace in the first draw?

```
P(A) = Number of ways it can happen/Total number of events
P(ACE) = 4/52 = 1/13
```

If you want to calculate the probability of drawing an ace in the second draw, the formula will change. Why? It's because the number of ace cards available changes from four to three.

Thus, if the first card chosen is an ace, the probability that the second card chosen is also an ace is called the *conditional probability* of drawing an ace. In this case, the *condition* is that the first card is an ace.

What is this probability? After an ace is drawn on the first draw, there are 3 aces out of 51 total cards left. This means that the probability that one of these aces will be drawn is 3/51 = 1/17.

If Events A and B are not independent, then you can use the following to figure it out:

```
P(A and B) = P(A) x P(B|A) = 4/52 x 3/51 = 1/221
```

Why Use Probability?

You can make better business decisions if you know the probability or chance of an event occurring. For example, a bank can set aside 2 percent as reserve funds if it knows that its loss ratio over the last two years was 2 percent. As another example, a store clerk can stock up on bills in denominations from $1 to $10 if he knows that 90 customers out of the last 100 customers have asked for change in denominations of $1 to $10.

These are simplistic uses of probability. Probability becomes more complicated when you start using probability distributions to make decisions. For example, the store owner plots the number of liters of milk that was demanded by customers over the last year. He finds that on 68 percent of the days the demand was for 50 liters of milk. For 2 percent of the days (holidays), there was demand for 100 liters of milk.

What is the amount of milk he should stock in his store? You can easily decide the following:

- On normal days he should stock 50 liters of milk.

- On holidays he should stock up to 100 liters of milk.

There are two major ways to calculate probability.

- Using Bayes' theorem, which explains conditional probability. This is a mathematical calculation.

- Using frequency to calculate probability.

  - For discrete variables, use histograms/frequency tables.

  - For continuous variables, use theorems related to distributions, especially the normal distribution. These include the empirical rule and Chebyshev's theorem.

Bayes' Theorem to Calculate Probability

There are two basic ideas that are related under Bayes' theorem probability and likelihood. When you toss a coin, you believe that the outcome of heads or tails will be 50/50. Such a coin is called a *fair* coin. In a *fraud* coin, the outcome will be biased toward heads or tails.

Suppose you have two coins and want to check which one is a fair coin. You toss both the coins ten times. The result is as follows: for Coin 1, you get five heads and five tails; for Coin 2, you get eight heads and two tails. You therefore conclude that Coin 1 is fair and Coin 2 is a biased coin.

Bayes' theorem explains how to update or revise the strengths of evidence-based beliefs on the basis of new evidence or a new occurrence. This is an important application because business scenarios keep changing and business managers need to make optimal decisions at each stage.

Bayes' theorem relates the conditional and marginal probabilities of stochastic events (that is, events that cannot be predicted exactly like the outcome of a toss of a coin). Let's call these Events A and B:

```
P(A|B) = (P(B|A)*P (A))/ P (B)
```

Each term in Bayes' theorem has a name:

- P(A) is the prior probability or marginal probability of A. It is *prior* in the sense that it does not take into account any information about B.

- P(A|B) is the conditional probability of A, given B. It is also called the *posterior* probability because it is derived from or depends upon the specified value of B.

- P(B|A) is the conditional probability of B, given A.

- P(B) is the prior or marginal probability of B and acts as a normalizing constant.

Bayes' Theorem in Terms of Likelihood

Let's look at this example:

```
P(A|B) α L(A|B)*P(A)
```

Here L(A|B) is the likelihood of A given a particular value of B. The rule is therefore a result of the relationship P(B|A) = L(A|B).

It has been seen that generally you can multiply the likelihood function L by a constant factor so that it is proportional to (though not equal to) the conditional probability P. I have already discussed that conditional probability P is the probability of Event A occuring when Event B occurs. For example, let's look at the chance of a rainbow coming if it rains; in this case, the conditional probability of Event A (the rainbow) will occur when Event B (the rain) occurs.

With this understanding, you can also state the following:

Posterior Probability (Probability of Event A to occur after Event B has occurred) = (Likelihood × Prior Probability) / Normalizing constant

As you can make out, the prior probability of Event A occurring is the probability *before* Event B has occurred.

The posterior probability is proportional to the product of the prior probability and the likelihood, as shown here:

```
Posterior = Normalized likelihood × Prior
```

Derivation of Bayes' Theorem from Conditional Probabilities

The probability of Event A given Event B has occurred is as follows:

```
P(A|B) = P (A∩B)/P(B)
```

Likewise, the probability of event B given event A has occurred is as follows:

```
P(B|A) = P (A∩B)/P(A)
```

Rearranging and combining these two equations, you find the following:

```
P(A|B) P(B) = P(A ∩ B) = P(B|A)/P(A)
```

Dividing both sides by P(B), providing that it is nonzero, you obtain Bayes' theorem (see Figure 6-1), as shown here:

```
P(A|B) = (P(B|A)*P(A))/P(B)
```

I = Running first mile in 4 mins
Z = Running second mile in 4 mins

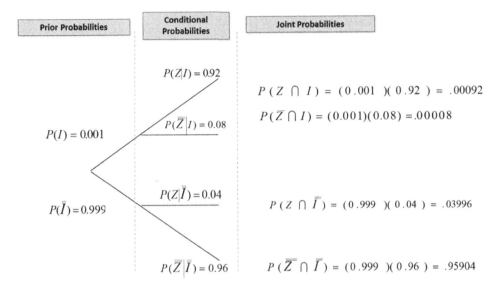

Figure 6-1. Bayes' theorem

In the Bayesian interpretation, probability means a degree of belief that an event will occur before and after evidence.

Thus, it is believed that a coin has a 50 percent chance of landing on heads or tails. One particular coin has a flaw. For that coin, after ten flips, you see that eight flips were heads and two were tails. In light of this evidence, the degree of belief that there is an equal chance of heads and tails will change for that particular coin. Thus, we can start talking about subjective probability—related to just one coin.

Decision Tree: Use It to Understand Bayes' Theorem

A simple trick to improve understanding of Bayes' theorem is to use a decision tree (see Figure 6-2). A decision tree serves the purpose of a graphical representation of Bayes' theorem and enables people to easily understand probabilities at different points of a process or a series of events. Every branch of the decision tree represents a possible decision or occurrence. The use of branches shows that each option is mutually exclusive. Because of the structure, it becomes easy to take a problem with many solutions and display it in a simple format that shows the relationships between different events and decisions.

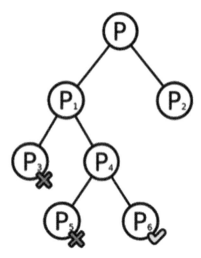

Figure 6-2. *Decision tree*

Decision trees provide a visual representation of choices, probabilities, and opportunities and is an easy way to break down complicated situations. The advantages are that it works like a flowchart and all parts are mutually exclusive.

You can use a decision tree to choose alternatives that

- Maximize the desirable outcome

- Minimize the undesirable outcome

- Are the most likely alternative among all the outcomes

Frequency to Calculate Probability

You will use the frequency of occurrence of an event (the number of times a particular event has occurred) to understand the chance/probability of the event occurring.

For Discrete Variables

For discrete variables, use histograms or frequency tables. These are two types of data representation that work best for discrete variables. You know that discrete variables are the variables in which you cannot have any decimal or in-between values. An example is the number of cars owned by a person.

In the frequentist interpretation, probability measures a proportion of outcomes. This is what you do when you look at pivot tables or frequency charts.

For Continuous Variables

For continuous variables, use theorems related to distributions, especially the normal distribution. These include the empirical rule and Chebyshev's theorem. You know that continuous variables can be whole numbers or decimal numbers. An example is the age of a customer.

Normal Distributions to Calculate Probability

For a discrete variable, the probability distribution contains the probability of each possible outcome, which is calculated by the count of the outcome divided by the total count. Thus, if you flip a coin ten times and get six heads and four tails, the probability distribution of heads is 60 percent and tails is 40 percent.

However, when you want to look at the distribution of probability of occurrence for a continuous variable (for example, time, age, income), then you will get a continuous curve because these variables can have many values, including decimal values. The distributions for continuous variables are called *continuous distributions' probability densities*. When this probability density curve is shaped like a bell, it is called the *normal distribution*. The most naturally occurring phenomena follow the normal distribution. If you were to plot the heights of adults in India, for example, you would see that there are few very short people (less than 4 feet) and few very tall people (greater than 7 feet); most people are somewhere in between (5 feet to 6 feet).

For a continuous variable with a sample size greater than 30 observations, the distribution approaches normality. Why 30? This number came as a result of simple sampling simulations from different parent populations (uniform, normal, exponential, triangular), and by the time the sample sizes reached 30 to 32, the distribution of the means started looking normal. That is the reason for the rule-of-thumb.

What are the characteristics of normal distributions?

- Normal distributions are symmetric around the mean.

- The mean and median of a normal distribution are equal.

- The area under the normal curve is equal to 1 or 100 percent.

- Normal distributions are denser in the center and less dense in the tails; they are bell-shaped.

- Normal distributions are defined by two parameters: the mean (μ) and the standard deviation (σ). Thus, you can generate a normal distribution sample if you put in the mean and standard deviation values.

■ **Note** The normal distribution is also called the Gaussian curve and the bell curve.

What do I mean by that last point? I mean that the standard normal distribution has certain characteristics. These are best understood under the empirical rule (also shown in Figure 6-3):

- 68 percent of the values/observations will lie between Mean+-1 standard deviation.

- 95 percent of the values/observations will lie between Mean+- 2 standard deviation.

- 99.7 percent of the values/observations will lie between Mean+- 3 standard deviation.

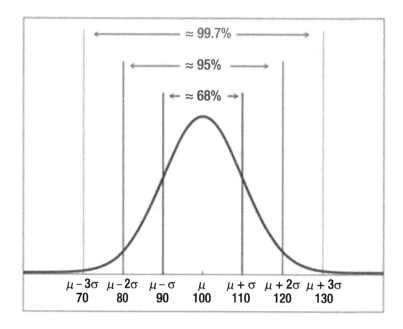

Figure 6-3. Normal distribution and probability as per the empirical rule

How can you calculate the probability of occurrence? You can use the empirical rule to understand the probability.

■ **Note**   The Six Sigma rule is an extension of the empirical rule. To achieve Six Sigma quality, a process must produce no more than 3.4 defects per million opportunities.

What If the Variable Is Not Normally Distributed?

What is the frequency of occurrence of events if the distribution type is not known? Or what if it can be seen that the distribution is not normal? The empirical rule applies only to bell-shaped distributions, and even then it is stated in terms of approximation. However, it is true that the mean and standard deviation define a distribution.

You can use Chebyshev's theorem to understand the minimum frequency that can be expected to live between the mean and standard deviation (also shown in Figure 6-4).

- At least 75 percent of the values/observations will lie between Mean+- 2 standard deviation.

- At least 88.89 percent of the values/observations will lie between Mean+- 3 standard deviation.

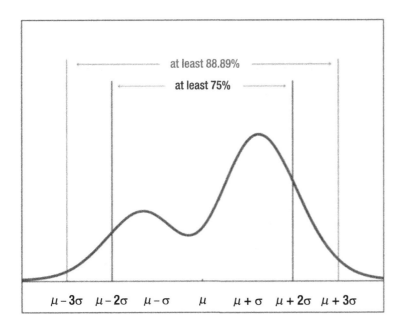

at least 88.89%

at least 75%

$\mu - 3\sigma$ $\mu - 2\sigma$ $\mu - \sigma$ μ $\mu + \sigma$ $\mu + 2\sigma$ $\mu + 3\sigma$

Figure 6-4. *Chebyshev's theorem*

Chebyshev's theorem gives the *minimum* proportion of the data that must lie within a given number of standard deviations of the mean. Thus, the actual proportions found within the indicated regions could be *greater* than what the theorem guarantees.

Case Study Using SAS

Let's look at using probability to solve a business problem using the SAS tool.

Problem Statement

A company has an automated service request system. All its customers can call into the help line number, and the service request is logged into the system by the customer care executive. An automated ticket ID or service request number (SR number) is generated. This ticket ID is shared with the customer. Once the request is successfully resolved, the customer service executive gives the customer a status call and closes the ticket on the system.

Other data points captured include the following:

- *Entitlement*: Type of request

- *Impact*: What the impact on the customer's business is because of the problem

- *Billable*: Whether customers are on a billable service plan or on a free service plan

- *Date Opened*: Time stamp when the service request was put in the system

- *Closed Date*: Time stamp when the service request was closed in the system

The company wants to understand the average time it takes to resolve queries. Does this average time vary by impact? What is the data that should be used to make a commitment on the resolution time to customers for different impacts?

Solution

Let's look at how to solve this business problem. You will use SAS for this.

The first step is to get the data into SAS. You point the SAS system to where the data is stored on your system or on the server.

```
/*SET WORKING DIRECTORY */

LIBNAME A "/home/subhashini1/my_content"; RUN;
```

You now import the data into the SAS system.

```
/*IMPORT DATA*/

FILENAME REFFILE "/home/subhashini1/my_content/Resolution time for Service request.csv"
TERMSTR=CR;

PROC IMPORT DATAFILE=REFFILE
    DBMS=CSV
    OUT=WORK.RESOLUTION;
    GETNAMES=YES;
RUN;
```

Once the data is in the system, you want to understand what variables are in the data and what formats the data variables are in.

```
/*CHECK CONTENTS OF THE DATA*/
PROC CONTENTS DATA=WORK.RESOLUTION ; RUN ;
```

Alphabetic List of Variables and Attributes				
# Variable	Type	Len	Format	Informat
5 Billable	Char	1	$1.	$1.
7 Closed Date	Num	8	DATETIME.	ANYDTDTM40.
6 Date Opened	Num	8	DATETIME.	ANYDTDTM40.
3 Entitlement	Char	55	$55.	$55.
4 Impact	Char	11	$11.	$11.
2 SR Type	Char	8	$8.	$8.
1 SR number	Char	13	$13.	$13.

Now you will use the DCOVA and I methodology to solve the project.

You will start by defining the business problem.

```
/* D = CREATE Y VARIABLE = CREATE RESOLUTON TIME */

DATA WORK.RESOLUTION;
SET WORK.RESOLUTION;
```

```
RESOLUTION_TIME = 'Closed Date'N-'Date Opened'N; RUN;

PROC UNIVARIATE DATA= WORK.RESOLUTION ;
VAR RESOLUTION_TIME; RUN ;
```

Missing Values			
Missing Value	Count	Percent Of	
		All Obs	Missing Obs
.	1	0.85	100.00

You can see that there is one row with a missing value. You therefore drop the row.

```
/* DROP 1 OBS WITH MISSING VALUE IN RESOLUTION TIME */


DATA WORK.RESOLUTION ;
SET WORK.RESOLUTION ;
WHERE RESOLUTION_TIME NE .; RUN ;

PROC MEANS DATA= WORK.RESOLUTION;
VAR RESOLUTION_TIME ; RUN ;
```

Analysis Variable : RESOLUTION_TIME				
N	Mean	Std Dev	Minimum	Maximum
117	4568414.36	5189604.59	6480.00	17142840.00

To make better sense of the data, you convert the resolution time. from seconds to days.

```
/* CONVERT RESOLUTION TIME INTO DAYS */


DATA WORK.RESOLUTION ;
SET WORK.RESOLUTION ;
RESOLUTION_DAYS= RESOLUTION_TIME /(24*60*60); RUN ;

PROC MEANS DATA= WORK.RESOLUTION;
VAR RESOLUTION_DAYS ; RUN ;
```

Analysis Variable : RESOLUTION_DAYS				
N	Mean	Std Dev	Minimum	Maximum
117	52.8751662	60.0648679	0.0750000	198.4125000

Let's look at some visualizations in a box plot.

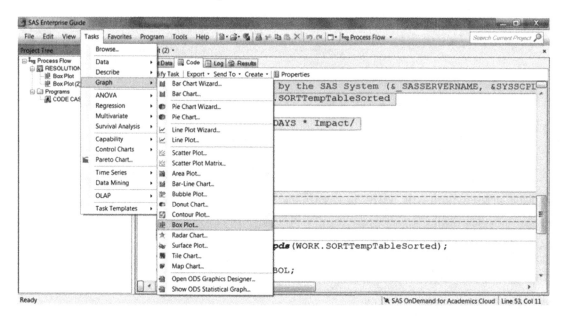

The code gets generated. As you know, when the button-driven menu is used to create graphs, you get the code as the SAS system auto-generates it. The auto-generated code looks like this:

```
/* ------------------------------------------------------------------
   Code generated by SAS Task

   Generated on: Friday, February 19, 2016 at 9:12:58 PM
   By task: Box Plot (2)

   Input Data: SASApp:WORK.RESOLUTION
   Server:   SASApp
   ------------------------------------------------------------------ */

%_eg_conditional_dropds(WORK.SORTTempTableSorted);
/* ------------------------------------------------------------------
   Sort data set SASApp:WORK.RESOLUTION
   ------------------------------------------------------------------ */

PROC SQL;
    CREATE VIEW WORK.SORTTempTableSorted AS
        SELECT T.Impact, T.RESOLUTION_DAYS
    FROM WORK.RESOLUTION as T
;
QUIT;
```

```
SYMBOL1      INTERPOL=BOX      VALUE=CIRCLE
    HEIGHT=1
    MODE=EXCLUDE
;
Axis1
    STYLE=1
    WIDTH=1
    MINOR=NONE

;
Axis2
    STYLE=1
    WIDTH=1
    MINOR=NONE

;
TITLE;
TITLE1 "Box Plot";
FOOTNOTE;
FOOTNOTE1 "Generated by the SAS System (&_SASSERVERNAME, &SYSSCPL) on
%TRIM(%QSYSFUNC(DATE(), NLDATE20.)) at %TRIM(%SYSFUNC(TIME(), TIMEAMPM12.))";
PROC GPLOT DATA=WORK.SORTTempTableSorted
;
    PLOT RESOLUTION_DAYS * Impact/
    VAXIS=AXIS1

    HAXIS=AXIS2

;
/* ----------------------------------------------------------------
    End of task code
    ---------------------------------------------------------------- */
RUN; QUIT;
%_eg_conditional_dropds(WORK.SORTTempTableSorted);
TITLE; FOOTNOTE;
GOPTIONS RESET = SYMBOL;
```

Box Plot

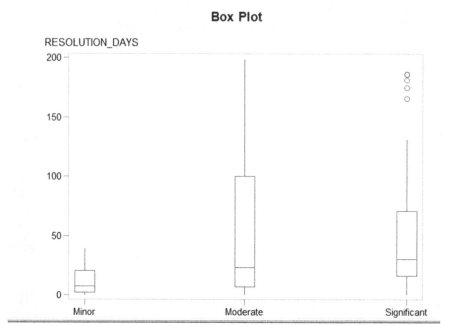

RESOLUTION_DAYS

___

■ **Note** Some service requests have a very long time to closure.

___

Let's look at the basic statistics about the variables in the data; these are called *descriptive statistics* and were explained earlier in the book.

```
/* CHECK DISTRIBUTION THRU TASKS > DESCRIBE > SUMMARY STATISTICS */
```

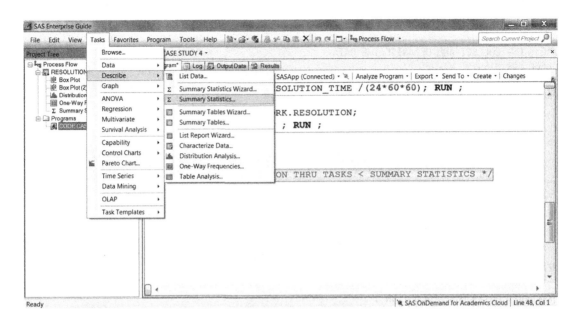

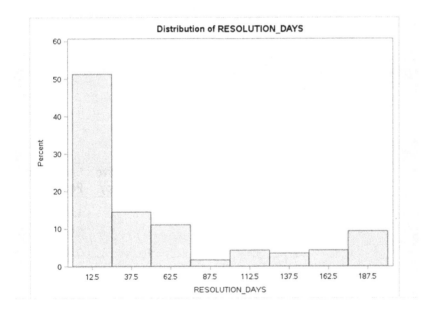

■ **Note**   Most of the values have a resolution time of less than or equal to 100 days.

Let's look at the next two stages of the DCOVA and I process: the collect and organize stages.
There is no data to merge or append, so there's nothing much to do in the collect stage.

Next you will take a look at the missing values as part of organizing the data.

```
/*C & O - COLLECT AND ORGANISE THE DATA
CHECK FOR MISSING VALUES */

PROC MEANS DATA=WORK.RESOLUTION; RUN ;
```

The MEANS Procedure

Variable	N	Mean	Std Dev	Minimum	Maximum
Date Opened	117	1727250303	13915441.23	1704272040	1751789880
Closed Date	117	1731818717	13973749.84	1705224600	1751884620
RESOLUTION_TIME	117	4568414.36	5189604.59	6480.00	17142840.00
RESOLUTION_DAYS	117	52.8751662	60.0648679	0.0750000	198.4125000

You can see missing values in non-numeric data by looking at PROC FREQ and not at PROC MEANS. PROC MEANS is used only for numeric data.

```
PROC FREQ DATA=WORK.RESOLUTION ;
TABLES Billable
Entitlement
Impact
'SR Type'N; RUN ;
```

Billable	Frequency	Percent	Cumulative Frequency	Cumulative Percent
N	107	91.45	107	91.45
Y	10	8.55	117	100.00

You can see the way the cases/observations are distributed between Billable y/n .

Entitlement	Frequency	Percent	Cumulative Frequency	Cumulative Percent
Operations & Alarm Management - BGP - 14	1	1.39	1	1.39
Preventive Maintenance - 1	1	1.39	2	2.78
Preventive Maintenance - 15	2	2.78	4	5.56
Preventive Maintenance - 3	1	1.39	5	6.94
Process History & Analytics - BGP - 7	1	1.39	6	8.33
Process Optimization - BGP	1	1.39	7	9.72
Process Optimization - BGP - 1	1	1.39	8	11.11
Process Optimization - BGP - 15	2	2.78	10	13.89
Process Optimization - BGP - 18	1	1.39	11	15.28
Process Optimization - BGP - 2	1	1.39	12	16.67
Process Optimization - BGP - 7	2	2.78	14	19.44
Requested Services - Fulfilment	14	19.44	28	38.89

Entitlement	Frequency	Percent	Cumulative Frequency	Cumulative Percent
Requested Services - Fulfilment - 1	1	1.39	29	40.28
Requested Services - Fulfilment - 2	7	9.72	36	50.00
Requested Services - Fulfilment - 3	4	5.56	40	55.56
Requested Services - Incident Support - 1	6	8.33	46	63.89
Requested Services Hiway Care Full - Fulfilment - 25	3	4.17	49	68.06
Requested Services SESP Basic - Fulfilment – 5	1	1.39	50	69.44
Requested Services SESP Basic - Fulfilment – 8	1	1.39	51	70.83
Requested Services SESP Basic - Incident Support - 5	1	1.39	52	72.22
Requested Services SESP Basic - Problem Management - 52	11	15.28	63	87.50
Requested Services SESP Remote - Fulfilment – 13	8	11.11	71	98.61
Requested Services SESP Remote - Incident Support - 5	1	1.39	72	100.00

Frequency Missing = 45

Impact	Frequency	Percent	Cumulative Frequency	Cumulative Percent
Minor	9	7.69	9	7.69
Moderate	63	53.85	72	61.54
Significant	45	38.46	117	100.00

SR Type	Frequency	Percent	Cumulative Frequency	Cumulative Percent
Incident	80	68.38	80	68.38
Problem	29	24.79	109	93.16
Request	8	6.84	117	100.00

▦ **Note** Forty-five observations are missing for Entitlement. All the other variables have no missing values.

No unique value variables can be used in the analysis since each observation will be different.

Let's look at duplicates in the SR number to see whether any observation has been repeated. If yes, you need to investigate why this is happening.

```
/* NO NEED TO CHECK SR number - WHICH IS THE PRIMARY KEY */
/*WE CAN CHECK FOR DUPLICATES IN PRIMARY KEY*/

PROC SORT DATA=WORK.RESOLUTION NODUPKEY OUT=WORK.RESOLUTION1;
BY 'SR number'N; RUN
```

Check the Log tab for details. You'll see the following output:

```
NOTE: There were 117 observations read from the data set WORK
NOTE: 0 observations with duplicate key values were deleted.
```

```
# OUTLIERS FOR CONTINUOUS VARIABLE -RESOLUTION TIME
/* OUTLIERS FOR RESOLUTION TIME - TOP 1% OF THE VALUES */
```

SAS Task to Do 1

As an exercise, you should write the PROC UNIVARIATE statement for the following output.

■ **Hint**   For which variable is the PROC UNIVARIATE statement written?

The UNIVARIATE Procedure
Variable: RESOLUTION_DAYS

Moments

N	117	Sum Weights	117
Mean	52.8751662	Sum Observations	6186.39444
Std Deviation	60.0648679	Variance	3607.78836
Skewness	1.26515175	Kurtosis	0.19770147
Uncorrected SS	745610.084	Corrected SS	418503.45
Coeff Variation	113.597502	Std Error Mean	5.552999

Basic Statistical Measures

Location		Variability	
Mean	52.87517	Std Deviation	60.06487
Median	23.68611	Variance	3608
Mode	.	Range	198.33750
		Interquartile Range	61.99722

Tests for Location: Mu0=0

Test	Statistic		p Value	
Student's t	t	9.521912	Pr > |t|	<.0001
Sign	M	58.5	Pr >= |M|	<.0001
Signed Rank	S	3451.5	Pr >= |S|	<.0001

Quantiles (Definition 5)

Level	Quantile
100% Max	198.41250
99%	191.93264
95%	185.97639
90%	167.46806

Quantiles (Definition 5)

Level	Quantile
75% Q3	70.97014
50% Median	23.68611
25% Q1	8.97292
10%	4.88819
5%	2.17847
1%	0.95000
0% Min	0.07500

Extreme Observations

Lowest		Highest	
Value	Obs	Value	Obs
0.075000	45	186.037	26
0.950000	96	187.228	103
0.979861	30	189.392	3
1.027083	51	191.933	35
1.096528	56	198.413	84

As part of organizing, let's remove any outliers or extreme values.

```
/* REMOVE VALUES GREATER THAN 191.93*/

DATA WORK.RESOLUTION ;
SET WORK.RESOLUTION ;
WHERE RESOLUTION_DAYS LE 191.93; RUN ;
```

Now the data is ready for the analyze phase in the DCOVA and I process. You will use the empirical rule and Chebychev's theorem.

```
/* A - ANALYSE ; USE EMERICAL RULE AND CHEBYCHEV'S THEOREM ;
NEED CHECK DISTRIBUTION BY IMPACT*/

PROC SORT DATA=WORK.RESOLUTION ;
BY IMPACT; RUN ;
PROC MEANS DATA=WORK.RESOLUTION N MEAN MEDIAN STDDEV MIN MAX;
BY IMPACT ;
VAR RESOLUTION_DAYS; RUN ;
```

The MEANS Procedure
Impact=Minor

Analysis Variable : RESOLUTION_DAYS

Maximum

39.1416667

    Impact=Moderate

Analysis Variable : RESOLUTION_DAYS

Maximum

189.3916667

    Impact=Significant

Analysis Variable : RESOLUTION_DAYS

Maximum

186.0368056

You want to look at the data across the subsegments.

```
/* CREATE SUBSETS ON TYPE OF IMPACT*/

DATA WORK.RESOLUTION_MINOR;
SET WORK.RESOLUTION;
WHERE IMPACT ='Minor'; RUN ;

DATA WORK.RESOLUTION_MOD;
SET WORK.RESOLUTION;
WHERE IMPACT ='Moderate'; RUN ;
```

Let's look at the next subsegment.

```
DATA WORK.RESOLUTION_SIG;
SET WORK.RESOLUTION;
WHERE IMPACT ='Significant'; RUN ;
```

You want to analyze the most common values and remove the values that occur rarely.

```
/* KEEP VALUES WHERE RESOLUTION_DAYS IS LE MEAN+2 SD*/

DATA WORK.RESOLUTION_MOD;
SET WORK.RESOLUTION_MOD;
WHERE RESOLUTION_DAYS LE (53.8889572+2*61.6009313); RUN ;
DATA WORK.RESOLUTION_SIG;
SET WORK.RESOLUTION_SIG;
WHERE RESOLUTION_DAYS LE (53.1293981+2*55.3275387); RUN ;

PROC MEANS DATA=WORK.RESOLUTION_MOD;
VAR RESOLUTION_DAYS; RUN ;
```

```
PROC MEANS DATA=WORK.RESOLUTION_SIG;
VAR RESOLUTION_DAYS; RUN ;
```

Let's compare the data sets across the basic statistics.

```
/* PROC MEANS FOR THE 3 SEGMENTS*/

PROC MEANS DATA=WORK.RESOLUTION_SIG;
VAR RESOLUTION_DAYS; RUN ;

PROC MEANS DATA=WORK.RESOLUTION_MOD;
VAR RESOLUTION_DAYS; RUN ;

PROC MEANS DATA=WORK.RESOLUTION_MINOR;
VAR RESOLUTION_DAYS; RUN ;
```

The MEANS Procedure

Analysis Variable : RESOLUTION_DAYS

Maximum

131.0465278

The MEANS Procedure

Analysis Variable : RESOLUTION_DAYS

N	Mean	Std Dev	Minimum	Maximum
56	42.1649802	49.3103093	0.9500000	167.4680556

The MEANS Procedure

Analysis Variable : RESOLUTION_DAYS

N	Mean	Std Dev	Minimum	Maximum
9	13.1111111	12.8386053	0.0750000	39.1416667

■ **Insight**   This is the last stage in the DCOVA and I process. What can you conclude from the analysis?

The average resolution time varies significantly by impact. The SLA can be derived using Chebychev's theorem and the empirical rule.

```
@ For Significant Impact cases
By Checbyshev's theorem
1. Atleast 75% of the cases get resolved between 0 - 106 days
2. Atleast 88.89% of the cases get resolved between 0-140 days
By Emperical Rule
1. 95% of the cases get resolved between 0 - 105.87 days
2. 99.97% of the cases get resolved between 0-140 days
```

```
@For Moderate Impact cases
By Checbyshev's theorem
1. Atleast 75% of the cases get resolved between 0 - 140 days
2. Atleast 88.89% of the cases get resolved between 0-190 days
By Emperical Rule
1. 95% of the cases get resolved between 0 - 140 days
2. 99.97% of the cases get resolved between 0-190 days
```

SAS Task to Do 2

As an exercise, you should define the significant ranges where the impact is minor.

Case Study in R

Let's see how to solve the same business problem using the R tool.

Problem Statement

A company has an automated service request system. All its customers can call into the help line number, and the service request is logged into the system by the customer care executive. An automated ticket ID or service request number (SR number) is generated. This ticket ID is shared with the customer. Once the request is successfully resolved, the customer service executive gives the customer a status call and closes the ticket on the system.

Other data points captured include the following:

- *Entitlement*: Type of request

- *Impact*: The impact on the customer's business because of the problem

- *Billable*: Whether customers are on a billable service plan or on a free service plan

- *Date Opened*: Time stamp when the service request was put in the system

- *Closed Date*: Time stamp when the service request was closed in the system

The company wants to understand the average time it takes to resolve queries. Does this average time vary by impact? What is the data that should be used to make a commitment on the resolution time to customers for different impacts?

Solution

Let's look at getting the data into the R tool. First set the location where the data is stored.

```
# SET WORKING directory
setwd('H://springer book//Case study//CaseStudy4')
```

Bring the data into the tool.

```
#IMPORT DATA FILE
Resolution<- read.csv("H:/springer book/Case study/CaseStudy4/Resolution time for Service
request.csv", stringsAsFactors=FALSE)
```

Let's see how the data looks using the str option.

```
#CHECK FORMAT

str(Resolution)

> str(Resolution)
'data.frame':    117 obs. of  7 variables:
 $ SR.number  : chr  "1-7657336422" "1-7658643852" "1-7735438423" "1-7880118403" ...
 $ SR.Type    : chr  "Incident" "Incident" "Incident" "Incident" ...
 $ Entitlement: chr  "Requested Services SESP Basic - Problem Management - 52"
                     "Requested Services SESP Basic - Problem Management - 52"
                     "Requested Services SESP Basic - Problem Management - 52"
                     "Requested Services SESP Basic - Problem Management - 52" ...
 $ Impact     : chr  "Moderate" "Significant" "Moderate" "Moderate" ...
 $ Billable   : chr  "N" "N" "N" "N" ...
 $ Date.Opened : chr  "9/11/2014 11:14" "9/11/2014 20:02" "10/8/2014 8:04" "11/13/2014 13:19" ...
 $ Closed.Date : chr  "10/16/2014 17:33" "10/22/2014 15:04" "4/15/2015 17:28" "12/19/2014 13:45" ...
```

Let's use the DCOVA and I methodology to solve the project.

```
Define the business problem and create the Y variable .

# D = CREATE Y VARIABLE = CREATE RESOLUTON TIME
# CONVERT Date.Opened AND Closed.Date INTO DATE TIME FORMAT

Resolution$Open <- as.Date(Resolution$Date.Opened, "%m/%d/%Y %H:%M")

Resolution$closed <- as.Date(Resolution$Closed.Date, "%m/%d/%Y %H:%M")
```

Use str code to understand whether the new variables created are as desired.

```
str(Resolution$Open)
str(Resolution$closed)

> str(Resolution$Open)
 Date[1:117], format: "2014-09-11" "2014-09-11" "2014-10-08" "2014-11-13" "2014-11-21" ...
> str(Resolution$closed)
 Date[1:117], format: "2014-10-16" "2014-10-22" "2015-04-15" "2014-12-19" "2014-12-07" ...

# DERIVED VARIABLE Y = RESOLUTION TIME

Resolution$resolution.time=(Resolution$closed -Resolution$Open)
View(Resolution)
str(Resolution$resolution.time)
> str(Resolution$resolution.time)
Class 'difftime'  atomic [1:117] 35 41 189 36 16 16 57 24 10 9 ...
  ..- attr(*, "units")= chr "days"
```

You can see that the format of the variables is character data. You need to convert it into numeric data.

```
# CONVERT RESOLUTION TIME INTO NUMERIC

Resolution$resolution.time= as.numeric(Resolution$resolution.time)
str(Resolution$resolution.time)
```

Let's look at creating a box plot to understand the distribution as well as look at outliers.

```
# Box plot
boxplot(Resolution$resolution.time, horizontal=TRUE, main="RESOLUTION_TIME")
```

RESOLUTION_TIME

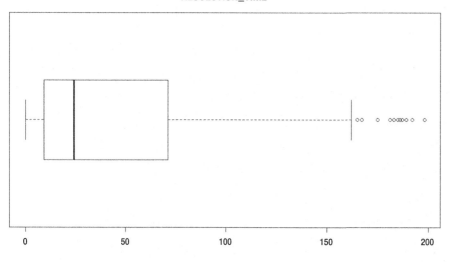

▪ **Note**   Some service requests have a very long time to closure.

Since the variable Resolution.time is a continuous variable, you can also look at a density plot to understand distribution.

```
# DENSITY PLOT
d<- density(Resolution$resolution.time)
plot(d)
```

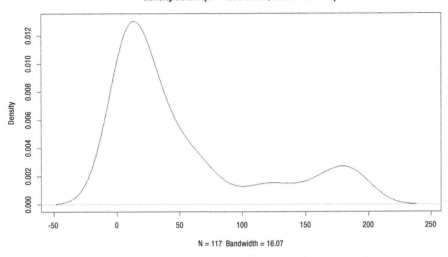

density.default(x = Resolution$resolution.time)

N = 117 Bandwidth = 16.07

■ **Note**   Most of the values have a resolution time of less than or equal to 100 days. Some values *may* have a resolution time of less than 0 days. Those are outliers and should be removed from the study.

```
# CHECK IF THERE ARE ANY VALUE LESS THAN 0 FOR RESOLUTION TIME
attach(Resolution)
Resolution$resolution.time.cat[resolution.time<0]<-0
Resolution$resolution.time.cat[resolution.time>=0]<-1

# CREATE TABLE TO CHECK FREQUENCY
mytable<- table(Resolution$resolution.time.cat)
mytable
```

View the table created.

```
> mytable
```

```
  1
117
```

Now let's look at the next stage in the DCOVA and I process: the collect and organize stages.

There is nothing to be done in the collect stage since there is no other data set you need to bring together. So, you start the organize phase with four components: missing values, outliers, dummy variables, and derived variables.

```
# C & O - COLLECT AND ORGANISE THE DATA

# CHECK FOR MISSING VALUES

#EXPLORE THE DATA TO UNDERSTAND NA AND OTHER SEGMENTS FOR DISCRETE VARIABLES
mytable<- table(Resolution$SR.Type)
mytable
```

```
> mytable
```

Incident	Problem	Request for Fulfillment
80	29	8

```
> mytable<- table(Resolution$Entitlement)
```

You can look at the table to check the missing values in this variable.

```
> mytable
```

	45
Operations & Alarm Management - BGP - 14	
	1
Preventive Maintenance - 1	
	1
Preventive Maintenance - 15	
	2
Preventive Maintenance - 3	
	1
Process History & Analytics - BGP - 7	
	1
Process Optimization - BGP	
	1
Process Optimization - BGP - 1	
	1
Process Optimization - BGP - 15	
	2
Process Optimization - BGP - 18	
	1
Process Optimization - BGP - 2	
	1
Process Optimization - BGP - 7	
	2
Requested Services - Fulfilment	
	14
Requested Services - Fulfilment - 1	
	1
Requested Services - Fulfilment - 2	
	7
Requested Services - Fulfilment - 3	
	4
Requested Services - Incident Support - 1	
	6
Requested Services Hiway Care Full - Fulfilment - 25	
	3
Requested Services SESP Basic - Fulfilment - 5	
	1
Requested Services SESP Basic - Fulfilment - 8	
	1
Requested Services SESP Basic - Incident Support - 5	
	1

```
Requested Services SESP Basic - Problem Management - 52
                                                    11
      Requested Services SESP Remote - Fulfilment - 13
                                                     8
  Requested Services SESP Remote - Incident Support - 5
                                                     1
```

You can do a similar check for the other variables.

```
> mytable<- table(Resolution$Impact)
> mytable

     Minor    Moderate Significant
         9          63          45
> mytable<- table(Resolution$Billable)
> mytable

  N    Y
107   10
```

▪ **Note**   There's no need to check for the primary key, which is the service request number. You cannot use a unique number in any model. It is used only to manage the data.

Now that you have finished with the missing values for categorical variables, you can look at the missing values for continuous variables.

```
# MISSING VALUES FOR CONTINUOUS DATA IN 10TH COLUMN- RESOLUTION TIME

Resolution[!complete.cases(resolution.time),10]

> Resolution[!complete.cases(resolution.time),10]
numeric(0)
```

The next treatment is for outliers, either extreme values or unusual values.

```
# OUTLIERS FOR CONTINUOUS VARIABLE -RESOLUTION TIME
# I HAVE DEFINED OUTLIERS AS BEING IN THE TOP .003% OF THE NORMAL DISTRIBUTION POPULATION
```

To detect and work on outliers, you will use a package called *outliers*.

```
install.packages("outliers")
library(outliers)

outs <- scores(Resolution$resolution.time, type="chisq", prob=0.997)
Resolution$resolution.time[outs]

> Resolution$resolution.time[outs]
numeric(0)

Note :- No outlier detected .
```

Now you can start the next phase in the DCOVA and I process: the analyze phase.

```
# A - ANALYZE :- FREQUENCY TABLE OF AVG. RESOLUTION TIME BY IMPACT
```

You will use a package called plyr for this.

```
library('plyr')

table1<- ddply(Resolution, c("Impact"), summarise,
          N    = length(resolution.time),
          mean = mean(resolution.time),
          median =median(resolution.time),
          sd   = sd(resolution.time)
          )
table1
```

```
> table1
       Impact  N    mean median       sd
1       Minor  9 13.11111      8 12.81059
2    Moderate 63 58.28571     24 65.47849
3 Significant 45 53.04444     30 55.27738
```

```
Note :- most of the cases fall under Moderate Impact . The Resolution time for Moderate
Impact is very high
```

```
# CREATE BOXPLOT TO CHECK DETAILS OF SEGMENTS UNDER IMPACT
```

You will create the box plot using the package ggplot2. Since the package is installed in your system, you just need to call the library code to activate it in the current session.

```
library(ggplot2)

bp1 <- ggplot(Resolution, aes(x=Impact, y=resolution.time)) +
  geom_boxplot()
bp1
```

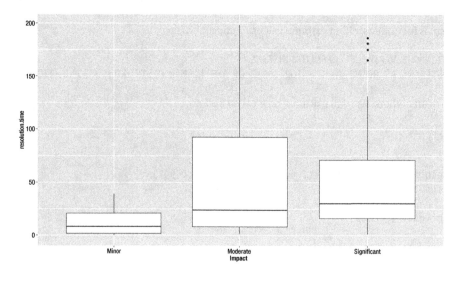

```
# PUT A DIAMOND WHERE THE MEAN IS
```

Say there are some small modifications you want to do such as putting a diamond where the mean value lies.

```
bp2<- ggplot(Resolution, aes(x=Impact, y=resolution.time)) + geom_boxplot() +
  stat_summary(fun.y=mean, geom="point", shape=5, size=4)
```

View the graph.

```
bp2
```

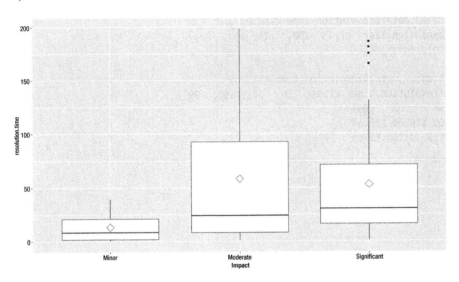

```
# FIND PERCENTILES
```

You also want to run some descriptive statistics for the data variable.

```
quantile(Resolution$resolution.time, c(.25, .50,  .75, .90, .99))

> quantile(Resolution$resolution.time, c(.25, .50,  .75, .90, .99))
   25%    50%    75%    90%    99%
  9.00  24.00  71.00 165.80 191.52

summary(Resolution$resolution.time)
> summary(Resolution$resolution.time)
   Min. 1st Qu.  Median    Mean 3rd Qu.    Max.
   0.00    9.00   24.00   52.79   71.00  198.00

# MAKE SUBSETS FOR VARIBLE IMPACT AND RUN QUARTILES
str(Resolution)

data1 <- subset(Resolution,Resolution$Impact=="Minor")
quantile(data1$resolution.time, c(.25, .50,  .75, .90, .99))
> quantile(data1$resolution.time, c(.25, .50,  .75, .90, .99))
```

```
  25%   50%   75%   90%   99%
 2.00  8.00 21.00 24.60 37.56

Data2 <- subset(Resolution,Resolution$Impact=="Significant")
quantile(data2$resolution.time, c(.25, .50,  .75, .90, .99))

> quantile(data1$resolution.time, c(.25, .50,  .75, .90, .99))
  25%   50%   75%   90%   99%
 16.0  30.0  71.0 151.4 186.0

# REMOVE OUTLIER FOR SIGNIFICANCE >99%

data2.1 <- subset(data2,data2$resolution.time <186.0)
quantile(data2.1$resolution.time, c(.25, .50,  .75, .90, .99))
boxplot(data2.1$resolution.time)

> data2.1 <- subset(data2,data2$resolution.time <186.0)
> quantile(data2.1$resolution.time, c(.25, .50,  .75, .90, .99))
   25%    50%    75%     90%     99%
 16.00  29.00  58.00  119.60  178.48
> boxplot(data2.1$resolution.time)
```

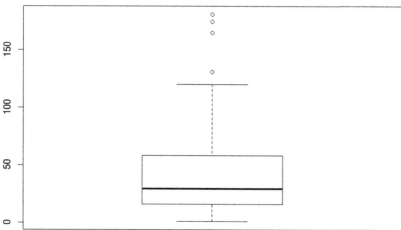

```
data1 <- subset(Resolution,Resolution$Impact=="Moderate")
quantile(data1$resolution.time, c(.25, .50,  .75, .90, .99))

> quantile(data1$resolution.time, c(.25, .50,  .75, .90, .99))
   25%    50%    75%     90%     99%
  8.00  24.00  92.50  178.20  194.28
```

You can draw some conclusions and make modifications in the data once you look at the output.

```
# REMOVE OUTLIER FOR MODERATE >99% AND REWORK BECAUSE MODERATE HAS A LONG TAIL IN THE BOXPLOT
```

You run dim to check how the data looks before modifications.

```
dim(data1)
data1.1 <- subset(data1,data1$resolution.time <=194.28)
quantile(data1.1$resolution.time, c(.25, .50,  .75, .90, .99))
boxplot(data1.1$resolution.time)

# REDO OUTLIER TREATMENT
data1.2 <- subset(data1.1,data1.1$resolution.time <=190.17)
quantile(data1.2$resolution.time, c(.25, .50,  .75, .90, .99))
boxplot(data1.2$resolution.time)

# REDO OUTLIER TREATMENT
data1.3 <- subset(data1.2,data1.2$resolution.time <=187.8)
quantile(data1.3$resolution.time, c(.25, .50,  .75, .90, .99))
boxplot(data1.3$resolution.time)
dim(data1.3)

# REDO OUTLIER TREATMENT
data1.4 <- subset(data1.3,data1.3$resolution.time <=185.82)
quantile(data1.4$resolution.time, c(.25, .50,  .75, .90, .99))
boxplot(data1.4$resolution.time)
dim(data1.4)

# REDO OUTLIER TREATMENT
data1.5 <- subset(data1.4,data1.4$resolution.time <=185.82)
quantile(data1.5$resolution.time, c(.25, .50,  .75, .90, .99))
boxplot(data1.5$resolution.time)
dim(data1.5)

> data1.5 <- subset(data1.4,data1.4$resolution.time <=185.82)
> quantile(data1.5$resolution.time, c(.25, .50,  .75, .90, .99))
   25%    50%    75%    90%    99%
  7.00  19.00  66.00 158.20 183.84
> boxplot(data1.5$resolution.time)
> dim(data1.5)
[1] 59 11
```

```
summary(data1.5$resolution.time)

> summary(data1.5$resolution.time)
   Min. 1st Qu.  Median    Mean 3rd Qu.    Max.
   1.00    7.00   19.00   49.25   66.00  185.00

sd(data1.5$resolution.time)
> sd(data1.5$resolution.time)
[1] 57.2284
```

■ **Insight**   This is the last phase of the DCOVA and I process.

The average resolution time varies significantly by impact.
The SLA can be derived using Chebychev's theorem for moderate impact cases.

- At least 75 percent of the cases get resolved between 0–163 days.

- At least 88.89 percent of the cases get resolved between 0–220 days.

And by empirical rule:

- 95 percent of the cases get resolved between 0–163 days.

- 99.97 percent of the cases get resolved between 0–220 days.

R Task to Do

As an exercise, you should define the significant ranges where the impact is minor and significant.

CHAPTER 7

■ ■ ■

Samples and Sampling Distributions Using SAS and R

In this chapter, you will look at sampling as a concept in statistical decision-making. I will cover sampling distributions and hypothesis testing.

Understanding Samples

What is a population? Any group with at least one common characteristic and is made up of people, transactions, products, and so on, is called a *population*. You need to understand the population for any project at the beginning of the project. In business, it is rare to have a population that has only one characteristic. Generally, it will have many variables in the data set.

What is a sample? A *sample* consists of a few observations or subset of a population. Can a sample have the same number of observations as a population? Yes, it can. Some of the differences between populations and samples are in the computations and nomenclatures associated with them.

In statistics, *population* refers to a collection of data related to people or events for which the analyst wants to make some inferences. It is not possible to examine every member in the population. Thus, if you take a *sample* that is random and large enough, you can use the information collected from the sample to make deductions about the population. For example, you can look at 100 students from a school (picked randomly) and make a fairly accurate judgment of the standard of English spoken in the school. Or you can look at the last 100 transactions on a web site and figure out fairly accurately the average time a customer spends on the web site.

The use of random numbers is generally preferred when trying to create a sample instead of taking every alternate or fifth specimen in the data set. Why is this? It's because there could be a regularity that is unforeseen in the population. Thus, a random sample is a better bet.

The other common way of sampling is called *stratified sampling*. Sometimes when there are distinct segments within the population, you will want to re-issue the sample to reflect the ratio in the population. For example, if you have 60 percent males in the organization, you will want the sample to reflect this. Thus, you would first divide the population into samples on the basis of gender and then for each gender type create a random sample.

In statistics, *precision* means that it is repeatable. An unbiased sample is one that is random. Note that as the sample size increases, the unbiased random sample gets more precise. This means if you were to repeat the statistics, you would arrive at a value that is very close to the original statistics. Thus, an increasing number of people from a population who are included in a sample will lead to the sample accurately representing the population provided a random process is used to build the sample. It is therefore true that if two or more samples are drawn from the same population, as they grow larger there is more likelihood

© Subhashini Sharma Tripathi 2016
S. S. Tripathi, *Learn Business Analytics in Six Steps Using SAS and R*, DOI 10.1007/978-1-4842-1001-7_7

that these two samples will resemble each other, especially if the random technique is followed. Thus, the differences or variations between two samples depend partly on the size of these two samples and the process of creating these two samples.

You will often hear that a sample of minimum size 30 is required for statistics. Note that this sample size of 30 is a rule of thumb only. I have found that you should look at least 30 observations for any one parameter or variable. Thus, if you want to measure the people who drink cola versus other drinks, you would look at 30 cola drinkers and 30 noncola drinkers. If, additionally, you have information about cola drinkers and noncola drinkers across four regions, you would want to look at 30 into 2 into 4. Thus, on average, it is good to go with sample sizes of 1,000 to 1,500, considering that business data has multiple variables and billions of observations.

What are the uses of sampling? In business, sampling enables you to indicate how much data to collect and how often it should be collected. This is especially true in the context of the test versus control experiments that you run.

Sampling becomes much less costly while investing in primary research. If you were trying to figure out which of the two types of burgers is liked best by the people of United States, you would want to look at interviews with 1,000 people as compared to the entire population of the United States.

Thus, if data collection is to be investigated afresh, sampling allows for greater speed and greater depth of data collection.

Is this relevant in the realm of business? It will depend on the department that you are considering in business. It is true that sampling for the purpose of data collection is more used in market research to understand a new customer segment or understand a reaction to a new policy or in research for academics.

You have seen that the probability theory is what helps you understand the occurrence of various events. Generally, if you use random sampling, you can be sure that you can come up with a good analysis on the population by looking at a sample. Here are some things you need to understand about a sample (part of a population):

- The probability of an event is the proportion of times that event occurs in a long series of experiments. Thus, it is the frequency of the event of interest occurring out of the total frequency of events.

- The probability of any event lies between 0 and 1. A probability equal to 0 is an impossible event, while a probability of an event equal to 1 means a sure event.

- The probability of an event of interest not occurring is equal to 1 minus the probability of the event of interest occurring.

- The probability of an event of interest occurring when data is selected at random (simple random sampling) is the same as the probability of the event occurring in the population. Thus, if 10 percent of the population is left-handed, the probability of left-handed people in the population is .1 or 10 percent, and there is a 10 percent chance that an individual chosen at random from this population will be left-handed.

Parameters are associated with population, and statistics are associated with samples. You compute descriptive statistics related to samples, and then you want to infer things related to populations. Thus, you want to prove that sample statistics are close to the parameters of the population.

There are various types of variables in a data set. When there are variables that can be represented by whole numbers and decimals, you call these variables *continuous variables*. For example, you can be 18 years old or 18.2 years old or 19 years old. Age is therefore a continuous variable.

There are some variables that can be represented only by whole numbers. For example, you can have one car or two cars, but not 1.5 cars. These types of variables are called *discrete variables*.

There are four levels of measurement.

- *Nominal* is represented by names. Examples are city, location, name of customer, and so on.

- *Ordinal* variables are those that have an order. Thus, you know that a metropolitan area has a bigger population than a city, which will have a bigger population than a town. An order is implied when a location is plotted as a metropolitan area, a city, and a town.

- *Intervals* are meaningful differences, by the way of slotting or segmenting. Thus, if someone gets more than 60 percent marks, he gets a first division, while if he is someone who gets between 50 to 60 percent marks, he gets a second division. These are meaningful intervals.

- *Ratio scale* is when a variable is represented as a ratio between 0 to 1. It could also be represented in percentages.

Sampling methodologies are divided into two basic types.

- *Probability sampling*: In probability sampling, each member of the population or sampling frame has a known probability of being selected, and this probability is greater than zero.

- *Nonprobability sampling*: In nonprobability sampling, members are selected from the population sampling frame in a nonrandom manner. It is quite possible that some members of the population in the sampling frame have zero probability of getting selected.

Nonprobability sampling includes the following:

- Self-selecting samples

- Convenience samples

- Judgmental samples

- Quota samples

As the names themselves show, these can all be very biased and may end up not representing the population correctly. You will not want to base your inferences on nonrandom sampling methodologies. Probability sampling methods include the following:

- *Simple random sampling* is where every member of the population has an equal likelihood of being selected. Thus, random sampling is the purest form of probability sampling.

- *Systematic sampling* is simple random sampling done in an ordered, systematic way; perhaps the population has been sorted in a certain order before the sampling starts. Once the required sample size has been calculated, its record is selected from the sampling frame. The assumption is that the list or sampling frame does not contain any hidden order (that is, no sorting, though often the data may be sorted on date of creation or date of transaction since that's intrinsic to the data).

- *Stratified sampling* is when the population is divided into different groups or stratas from which you can sample. This is specific to certain statistical analysis techniques such as logistic regression where you want a 50-50 sample of goods and bads. Generally, it is supposed to reduce sampling error. A strata is a subset of the population that shares some common characteristic, such as male or female, high net worth, and so on. The first step is to identify the strata and their actual representation in a percentage of the actual population. After this, random sampling is used to select members from each strata of the sampling frame. Each strata in the sample will be presented in a similar ratio as it is represented in a population.

- *Cluster sampling* is when the population is divided into clusters. Some clusters will be chosen at random, and within the cluster the units are chosen by the simple random sampling methodology. So, if you want to study the behavior of populations, you may divide the population into clusters like puppies (poor urban professionals) and woofs (well-off older folks).

It is quite possible that you do not get access to the entire population to create a sample. The list of subjects from which a sample is selected is called the *sampling frame*. This could mean you have only the sensors database, employee database, customer database, and so on.

Even among the random sampling methodologies, cluster sampling and stratified sampling are often used to solve the same problem. Cluster sampling is easier and cheaper, especially if data is not available in-house and you are going to do some data collection. Stratified sampling is easier and used more when the data is already present and you want to treat subgroups in a particular manner.

Sampling Distributions

What will happen if you are able to draw out all possible samples of 30 or more observations from a given population/sample frame? For each of these samples, you could compute the descriptive statistics (mean, median, standard deviation, minimum, maximum). Now if you were to create a probability distribution of this statistic, it would be called the *sampling distribution*, and the standard deviation of this statistic would be called the *standard error*.

Suppose you sample 100 people from a population of Delhi between the ages of 20 to 30 and calculate the descriptive statistics (mean, median, mode, minimum standard deviation, maximum standard deviation) for the set of people. Will you expect that the result of the sample should be equal to or similar to the total population of Delhi? It might be marginally lower or marginally higher, but overall you should see great synergy.

If you took another sample of 100 people from the same population of Delhi, how do you think the descriptive statistics of this second population would look compared to the first? Would it not be similar?

I already spoke about the fact that inferential statistics is concerned with being able to make conclusions about a population from a sample. You also saw that this is generally done with random sampling, and then by looking at the sample and the descriptive statistics of the sample, you make conclusions about other samples or about the entire population.

Thus, to make an inference, you should be able to understand how the sample statistics is likely to marry between multiple samples taken from the same population/sample frame.

It has been found that if infinite numbers of samples are taken from the same sample frame/population and a sample statistic (say, the mean of the samples) is plotted out, you will find that a normal distribution emerges. Thus, most of the means will be clustered around the mean of the sample mean, which incidentally will coincide or be very close to the population/sample frame mean. This is as per the normal distribution rule, which states that values are concentrated around the mean and few values will be far away from the mean (very low or very high as compared to the mean).

I am talking about the mean as a statistic here. You can also consider any other value from the descriptive statistics family and find the same thing.

I have taken a sample of 14 means and created a `Sample_mean_1` histogram for a sampling frame/population and plotted it (see Figure 7-1). Let's see how it looks.

Variable	Observations	Obs. with missing data	Obs. without missing data	Minimum	Maximum	Mean	Std. deviation
Sample_m	14	0	14	-225.399	500.734	65.689	195.694

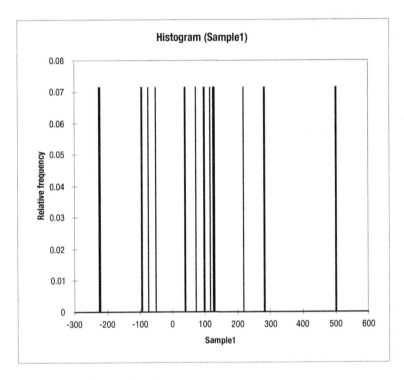

Figure 7-1. *The Sample1 histogram*

You can see that the mean doesn't show much variation in the frequency of occurrence. Now when I do this for a larger sample of means, let's see what happens (see Figure 7-2).

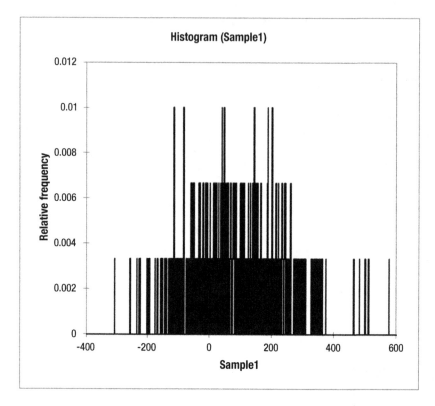

Figure 7-2. *The updated histogram*

A normal distribution is emerging. I will plot at this out in a normal distribution to show how it looks (Figure 7-3).

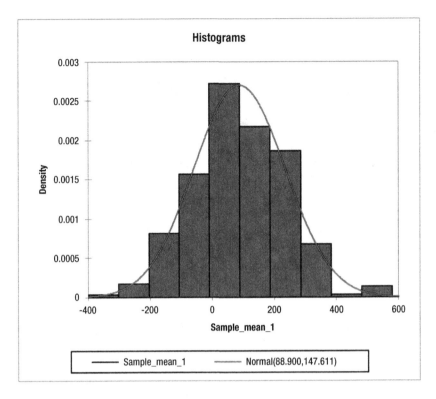

Figure 7-3. *Looking at the normal distribution*

As the number of data points increases, you can see that the distribution becomes normal. This is interesting because this is the outcome whatever the distribution of the sampling frame population.

Different shapes that show up in a histogram are called different *distributions*. The common type of distributions are as follows.

Discrete Uniform Distribution

This is when every outcome has the same frequency (Figure 7-4).

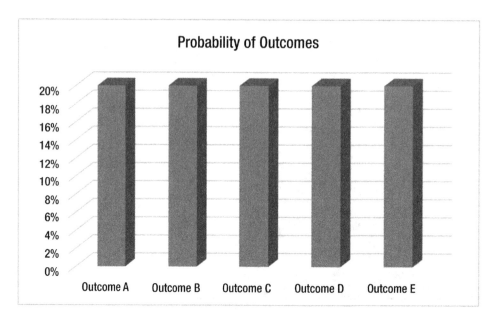

Figure 7-4. *A discrete uniform distribution*

Binomial Distribution

The basic building block of the binomial distribution is a Bernoulli random variable. This is a variable for which there can be only two possible outcomes, and the probability of these outcomes satisfies the conditions of a valid probability distribution function, which is that each probability is between 0 and 1 and the total probabilities sum up to 1 or 100 percent.

Since a single observation of the outcome of a burn on a random variable is called a *trial*, the sum of a series of such trials is distributed as a binomial distribution.

Thus, one such example is the probability of getting a tail on the toss of a coin, which is 50 percent or .5. If there are 100 such tosses, you will find that getting 0 heads and 100 tails is very unlikely, getting 50 heads and 50 tails is the most likely, and getting 100 tails and 0 heads is the most unlikely.

Now let's look at a scenario where you have four possible outcomes and the probability of getting outcome 1, 2, or 3 defines success, while getting an outcome of 4 defines failure. Thus, the probability of success is 75 percent, and the probability of failure is 25 percent. Now if you were to try 200 tosses again, you will find that a similar distribution occurs, but the distribution will be more skewed or can be seen to be a bit shifted as compared to the earlier 50-50 distribution (Figure 7-5).

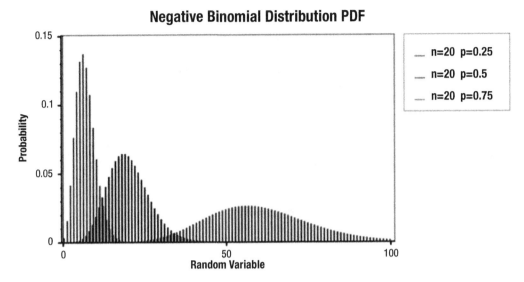

Figure 7-5. *A demonstration of binomial distribution*

Continuous Uniform Distribution

What if you have no prior beliefs about the distribution of probability or if you believe that every outcome is equally possible? It's easier when the value is discrete for a variable. When this same condition is seen over a continuous variable, the distribution that emerges is called the *continuous uniform distribution* (Figure 7-6). It is often used for random number generation in simulations.

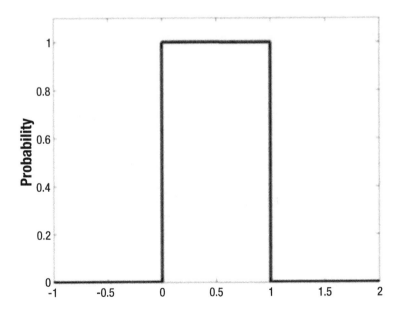

Figure 7-6. *A continuous uniform distribution*

Possion Distribution

Let's look at some events that occur at a continuous rate, such as phone calls coming into a call center. Let the rate of occurrence be *r*, or lambda. When the number is small (that is, there is only one or two calls in a day), the possibilities that you will get zero calls on certain days is high. However, say the number of calls in the call center is on average 100 per day. Then the possibility that you will ever get zero calls in a day is very low. This distribution is called the Poisson distribution (Figure 7-7).

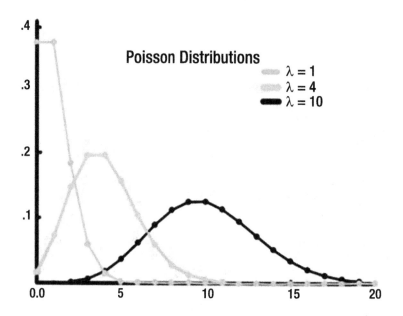

Figure 7-7. *Some Poisson distributions*

Use of Probability Distributions

Sales probability distributions are statistical models that show the possible outcome of a particular event or the probability of an outcome of an event. They are used extensively for the following:

- *Scenario analysis*: You can use them to create business scenarios and understanding the probability of certain outcomes. Thus, you can figure out the worst-case, likeliest, and best-case outcomes.

- *Business and sales forecasting*: You can predict future levels of sales and plan for future events. Thus, the company can create basic business plans on some scenarios but still be aware of other possible scenarios.

- *Risk evaluation*: You can conduct stress testing to understand the capability to withstand different types of risk by the business based on probability distributions.

Central Limit Theorem

The *central limit theorem* states that the sampling distribution of the mean of any independent random variable will be normal or nearly normal if the sample size is large enough. How much is large enough?

- *How closely do you want the sampling distribution to resemble a normal distribution?* If you want it to be nearly normal, then you need more samples as compared to if you want it to be less nearly normal.

- *What is the shape of the underlying distribution?* If you pick out samples from a population that is originally normally distributed, there is a higher chance that with a few sample points you will be able to get a normal distribution for the mean of the samples.

However, the rule of thumb and practice in the industry is that for one variable, a sample size of 30 is large enough if the original population is nearly normal and bell-shaped. However, if the original population is distinctly not normal (that is, it has outliers, multiple peaks, and so on), then you might need to pick out a sample size that is much larger.

To summarize, the central limit theorem states that the distribution of a large number of samples taken from the same population are approximately normal irrespective of how the distribution of the population is. The central limit theorem makes inferencing much easier using statistics.

The Law of Large Numbers

In the probability theory, there is a theorem called the *law of large numbers*. This theorem states that if the average values are obtained from a number of trials or instances of any event, the average result will be close to the expected value and will become even closer to this expected value as the count goes up. Thus, the law of large numbers guarantees stable, long-term results for the average value of certain events.

In other words, in statistics, as the number of similarly generated values across one variable increases, the sample mean approaches the theoretical mean. This was proven by Bernoulli. He considered a game of chance with only two outcomes: win or loss. He considered the game across a very large sample of plays and players and found that as the sample of plays and players increased, the probability that someone would win or lose increasingly became 50 percent.

If this example was to be seen for a roll of dice where the probability of each outcome is 1/6, it will be seen that as the role of dice increases beyond 100 rolls, this one in six (1/6) probability is maintained on an even keel.

Thus, the central limit theorem helps convert any type of distribution into a normal distribution for the purposes of inferencing. You are able to make smart deductions and conclusions on the basis of the central limit theorem. Thus, the process is, irrespective of any distribution that exists in the population, as follows:

- You should take many samples.

- If you plot the means of this many samples, the probability distribution of the means will start resembling a normal distribution.

- All the previously mentioned inferencing theorems such as the empirical rule will therefore apply to this set of samples.

Why do you use statistics for decision-making in business?

One of the reasons is that it is usually not a practical proposition to work on the entire population. Instead, you would want to create a sample of the population and then use statistics from that random sample to draw some conclusions that will apply to the population. Many of the common statistical procedures have a requirement that the data should be approximately or nearly normal. That is, it should follow the normal distribution.

Then what do you do when the population is not normal? That's when the central limit theorem helps. The central limit theorem says that if you have a sufficiently large number of randomly selected independent samples or a sufficiently large number of randomly selected independent observations, the means of these samples will follow a normal distribution. This is true even if the base population from which the sample is coming does not follow the normal distribution.

The next question becomes how can you figure out which sample is a good sample? In other words, how can you ensure it adequately represents the population and can be used to make inferences?

Also, if you have two samples, how can you conclude that these two samples are similar or dissimilar? The underlying question is, do these samples come from the same population?

The hypothesis test is a statistical test that is used to answer the question, is there enough evidence to infer that a certain condition, which is true for the sample, is true for the entire population?

There can obviously be two hypothesis related to this.

- The *null hypothesis* is the statement that the means of the two samples or the sample and the population are the same.

- The *alternate hypothesis* is that the null hypothesis is not true.

The test has to determine whether to reject the null hypothesis or to accept the null hypothesis and reject the alternate hypothesis. What is a hypothesis?

A hypothesis is an assumption or supposition about a population parameter.

So, you can assume that the average height of Indian women is 5'6". The null hypothesis for the problem would be that the average height of an Indian woman is 5'6". The alternate hypothesis would be that the average height of an Indian woman is not 5'6" (it is either more or less than 5'6").

The outcome of the test will tell you whether you can reject the null hypothesis or you have failed to reject the null. Failure to reject the null implies that the data is not sufficient or robust enough for you to reject the null.

The two families of tests depend on the type of data.

- Tests for parametric data

- Tests were nonparametric data

The assumption for parametric data is that the variables under consideration are normally or nearly normally distributed. Thus, parametric tests can perform well with skewed and normal distributions if the sample size is large enough.

■ **Note** Nonparametric tests also perform well when the spread of data (minimum to maximum) is different for different groups of the same variable.

Thus, to summarize, for parametric tests, the distribution is assumed to be normal, and the variance is assumed to be homogeneous. The typical data would be ratio interval scale data, and the data set would be adequately represented by the mean as a measure of central tendency. Generally, you can draw more conclusions with parametric tests.

Nonparametric tests are used when there is no assumption related to the distribution. There is no assumption of homogeneity about the variance when the data is ordinal or nominal (for example, customer service satisfaction score data), and median is a much better central tendency to understand the data as compared to the mean. This type of data is usually less affected by outliers since the possible categories are limited for the variable under consideration.

Nonparametric tests are generally less powerful and often used for small data sizes and ordinal nominal data. It is good practice to use parametric tests as much as possible. You can test for equal variance, usually with an f-test.

An f-test is designed to test whether the population variance is equal. It is thus a test that can be done before choosing a parametric order nonparametric test. It does this by comparing the reissue of two variances, so if the ratio is 1 or nearly 1, you can assume that the variances are nearly equal.

Parametric Tests

The following are some parametric tests:

- *Students t-test*: Student t-tests look at the differences between two groups across the same variable of interest. Or they look at two variables in the same sample. The consideration is that there can be only two groups at maximum.

- An example is if you want to compare the grades in English for students of Class 1's Section A and Section B. Another example is if you want to compare the grades in Class 1's Section A for math and for science.

  - *One sample t-test*: When the null hypothesis reads that the mean of a variable is less than or equal to a specific value, then that test is one sample t-test.

  - *Paired sample t-test*: When the null hypothesis assumes that the mean of variable 1 is equal to the mean of variable 2, then that test is a paired sample t-test.

  - *Independent sample t-test*: This compares the mean difference between two independent groups for a given variable. The null hypothesis is that the mean for the variable in sample 1 is equal to the mean for the same variable in sample 2. The assumption is that the variance or standard deviation across the samples is nearly equal.

For example, if you want to compare the grades in English for students of Class 1's Section A and Section B, you can use an analysis of variance (ANOVA) test as a substitute for the students' t-test.

- *ANOVA test*: This is the significance of differences between two or more groups across one or more categorical variable. Thus, you will be able to figure out whether there is a difference between groups, which is significant, but it will not tell you which group is different.

- An example is if you want to compare the grades in English for students of Class 1's Section A, Section B, and Section C. Another example is if you want to compare the grades in Class 1's Section A for math, English, and science.

  - *One-way ANOVA*: In this test, you compare the mean of a number of groups based on one independent variable. There are some assumptions like that the dependent variable is normally distributed and that the group of independent variable groups have equal variance on the dependent variable.

  - *Two-way ANOVA*: Here you can look at multiple groups and two variables of factors. Again, the assumption is that there is homogeneity of variance and the standard deviation of the population of all the groups are similar.

Nonparametric Tests

Here the data is not normally distributed. Thus, if the data is better represented by the median instead of the mean, it is better to use nonparametric tests. It is also better to use nonparametric tests. If the data sample is small, the data is ordinal ranked or may have some outliers that you do not want to remove.

Chi-squared tests compare observed frequencies to expected frequencies and are used across categorical variables.

As discussed, chi-square tests will be used on data that has ordinal nominal variables. For example, say you want to understand the population of Indian males in cities who regularly exercise, sporadically exercise, or have not exercised over the last 20 years. Thus, you have three responses tracked over 20 years, and you need to figure out whether the population has shifted between year 1 and year 20. The null hypothesis here would mean that there is no change or no difference in the situation.

- *Year 1 statistics*: 60 percent regularly exercise, 20 percent sporadically exercise, and 20 percent have not exercised.

- *Year 20 statistics*: 68 percent regularly exercise, 16 percent sporadically exercise, and 16 percent have not exercised.

The test for both years was run on 500 people. Now you would compare the year 20 statistics with what could be the expected frequencies of these people in year 20 (if the year 1 trends are followed) as compared to the observed frequencies.

Case Study Using SAS

A magazine does a survey and wants to publish the results. The following data represents the business startup costs (in thousands of dollars) for shops, gathered through a survey. One of the questions to be answered is, is it equally expensive to start any of the following businesses or are some businesses cheaper to start?

Here are descriptions of the variables:

- X1 = Startup costs for pizza restaurant

- X2 = Startup costs for bakery

- X3 = Startup costs for shoe store

- X4 = Startup costs for gift shop

- X5 = Startup costs for pet store

What is the conclusion that the article should publish?

Let's start with define, the first stage in the DCOVA and I process.

```
/*D - Define the business Y - check if the distribution of startup costs are similar across
lines of business*/
```

You need to get the data set into the SAS tool.

```
/* IMPORT DATA*/

DATA WORK.STARTUP;
    LENGTH
        X1              8
        X2              8
        X3              8
        X4              8
        X5              8 ;
    FORMAT
        X1              BEST3.
        X2              BEST3.
        X3              BEST3.
        X4              BEST3.
        X5              BEST3. ;
    INFORMAT
        X1              BEST3.
        X2              BEST3.
        X3              BEST3.
        X4              BEST3.
        X5              BEST3. ;
    INFILE '/home/subhashini1/my_content/startupcost.csv'
        DLM=','
        MISSOVER
        DSD ;
    INPUT
        X1              : ?? BEST3.
        X2              : ?? BEST3.
        X3              : ?? BEST3.
        X4              : ?? BEST3.
        X5              : ?? BEST3. ;
RUN;
```

How does the data look? Let's understand the data.

```
/*UNDERSTAND THE DATA */

PROC CONTENTS DATA = WORK.STARTUP; RUN ;

PROC PRINT DATA=WORK.STARTUP (OBS= 10) ; RUN ;
```

You can conduct a PROC UNIVARIATE to know more about the numerical variables.

```
PROC UNIVARIATE DATA= WORK.STARTUP NORMAL PLOT;
VAR  X1 X2 X3 X4;
HISTOGRAM X2 X3 X4 / NORMAL ; RUN ;
```

As you are using the SAS tool, you will need to structure the data so that the tool can work well.

```
/* STRUCTURE THE DATA FOR T TEST X1 VS X2 */

DATA WORK.STARTUP_X1;
SET WORK.STARTUP;
KEEP X1; RUN ;

DATA WORK.STARTUP_X1;
SET  WORK.STARTUP_X1;
SEGMENT = 1;
RENAME X1 = VALUE;RUN ;

PROC PRINT DATA=WORK.STARTUP_X1 (OBS=10) ; RUN ;

DATA WORK.STARTUP_X2;
SET WORK.STARTUP;
KEEP X2;
 RUN ;

DATA WORK.STARTUP_X2;
SET  WORK.STARTUP_X2;
SEGMENT = 2;
RENAME X2 = VALUE;
RUN ;

PROC PRINT DATA=WORK.STARTUP_X2 (OBS=10) ; RUN ;

DATA TOTAL_X1_X2;
SET WORK.STARTUP_X1 WORK.STARTUP_X2;RUN ;
```

Now you sort the data.

```
PROC SORT DATA= TOTAL_X1_X2;
BY SEGMENT; RUN ;
```

There is no process to accomplish in the collect phase.
You can now start the organize phase. You will work on the missing values.

```
/* UNDERSTAND MISSING VALUES */

PROC MEANS DATA=TOTAL_X1_X2 N NMISS MEAN STDDEV MEDIAN MIN MAX;
CLASS SEGMENT ; RUN ;

/* REMOVE MISSING VALUES*/

DATA FINAL_X1_X2;
SET TOTAL_X1_X2;
WHERE VALUE NE .; RUN ;
PROC MEANS DATA=FINAL_X1_X2 N NMISS MEAN STDDEV MEDIAN MIN MAX;
CLASS SEGMENT ; RUN ;
```

Let's understand if the variables are normally distributed.

```
/* CHECK NORMALITY */

PROC UNIVARIATE DATA=FINAL_X1_X2 NORMAL;
QQPLOT VALUE / NORMAL (mu=est sigma=est color=red l=1);
BY SEGMENT; RUN ;
```

The output looks as follows:
The UNIVARIATE Procedure
SEGMENT=1

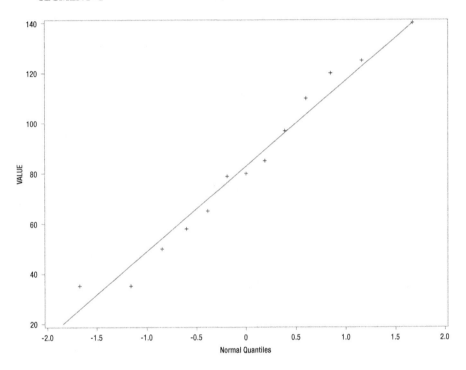

The UNIVARIATE Procedure
Variable: VALUE
SEGMENT=2

Moments

N	22	Sum Weights	22
Mean	92.0909091	Sum Observations	2026
Std Deviation	37.956001	Variance	1440.65801
Skewness	0.47019131	Kurtosis	-0.6371584
Uncorrected SS	216830	Corrected SS	30253.8182
Coeff Variation	41.2157957	Std Error Mean	8.09224659

Basic Statistical Measures

Location		Variability	
Mean	92.09091	Std Deviation	37.95600
Median	87.00000	Variance	1441
Mode	40.00000	Range	120.00000
		Interquartile Range	60.00000

Note: The mode displayed is the smallest of 11 modes with a count of 2.

Tests for Location:Mu0=0

Test	Statistic		p Value	
Student's t	t	11.38014	Pr > |t|	<.0001
Sign	M	11	Pr >= |M|	<.0001
Signed Rank	S	126.5	Pr >= |S|	<.0001

Tests for Normality

Test	Statistic		p Value	
Shapiro-Wilk	W	0.921498	Pr < W	0.0817
Kolmogorov-Smirnov	D	0.158329	Pr > D	>0.1500
Cramer-von Mises	W-Sq	0.084843	Pr > W-Sq	0.1749
Anderson-Darling	A-Sq	0.572258	Pr > A-Sq	0.1261

Quantiles(Definition 5)

Level	Quantile
100% Max	160
99%	160
95%	160
90%	150
75% Q3	120
50% Median	87
25% Q1	60
10%	45
5%	40
1%	40
0% Min	40

Extreme Observations

Lowest		Highest	
Value	Obs	Value	Obs
40	26	120	27
40	15	150	14
45	31	150	25
45	20	160	18
60	30	160	29

The UNIVARIATE Procedure

SEGMENT=2

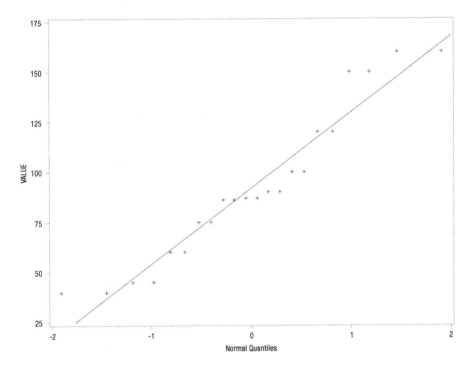

Conclusion: The QQplots appear to be linear. The null hypothesis for the normality test is that there is there is no significant departure from normality. Since the p-value is greater than .05, you cannot reject the null.

■ **Note**   If the sample size is greater than 2000, you should use the Kolmgorov test. If the sample size is less than 2000, the Shapiro test is better.

Here's a problem for you to practice: do this for the other variables in the data set and make a conclusion on the distribution.

Now you can start on the analyze phase of DCOVA and I.

Let's start by running the t-tests.

```
/* RUN T TEST FOR 2 SAMPLES, INDEPENDENT GROUP */

PROC TTEST DATA=FINAL_X1_X2;
CLASS SEGMENT;
VAR VALUE; RUN ;
```

The output looks as follows :-

The TTEST Procedure
Variable: VALUE

SEGMENT	N	Mean	Std Dev	Std Err	Minimum	Maximum
1	13	83.0000	34.1345	9.4672	35.0000	140.0
2	22	92.0909	37.9560	8.0922	40.0000	160.0
Diff (1-2)		-9.0909	36.6126	12.8080		

SEGMENT	Method	Mean	95% CL Mean		Std Dev	95% CL Std Dev	
1		83.0000	62.3727	103.6	34.1345	24.4774	56.3471
2		92.0909	75.2622	108.9	37.9560	29.2015	54.2416
Diff (1-2)	Pooled	-9.0909	-35.1490	16.9671	36.6126	29.5308	48.1923
Diff (1-2)	Satterthwaite	-9.0909	-34.6218	16.4400			

Method	Variances	DF	t Value	Pr > |t|
Pooled	Equal	33	-0.71	0.4828
Satterthwaite	Unequal	27.54	-0.73	0.4716

Equality of Variances

Method	Num DF	Den DF	F Value	Pr>F
Folded F	21	12	1.24	0.7210

Conclusion: Since the p-value is greater than .05, you *cannot* reject the null that mean 1 = mean 2.

Here's a problem for you to practice: do this for the other variables in the data set and make a conclusion about the hypothesis test.

Let's run the ANOVA process.

```
/* RUN ANOVA (IN ANOVA THE ASSUMPTIONS ARE TESTED AFTER RUNNING THE MODEL)*/

PROC ANOVA DATA=WORK.FINAL_X1_X2;
CLASS SEGMENT;
MODEL VALUE = SEGMENT ;
MEANS SEGMENT /HOVTEST=LEVENE;RUN;
```

The output looks as follows:
The ANOVA Procedure

Class Level Information

Class	Levels	Values
SEGMENT	2	1 2

Number of Observations Read	35
Number of Observations Used	35

The ANOVA Procedure
Dependent Variable: VALUE

Source	DF	Sum of Squares	Mean Square	F Value	Pr > F
Model	1	675.32468	675.32468	0.50	0.4828
Error	33	44235.81818	1340.47934		
Corrected Total	34	44911.14286			

R-Square	Coeff Var	Root MSE	VALUE Mean
0.015037	41.27019	36.61256	88.71429

Source	DF	Anova SS	Mean Square	F Value	Pr > F
SEGMENT	1	675.3246753	675.3246753	0.50	0.4828

The ANOVA Procedure

Levene's Test for Homogeneity of VALUE Variance ANOVA of Squared Deviations from Group Means

Source	DF	Sum of Squares	Mean Square	F Value	Pr > F
SEGMENT	1	733641	733641	0.37	0.5455
Error	33	64871914	1965816		

Welch's ANOVA for VALUE

Source	DF	F Value	Pr > F
SEGMENT	1.0000	0.53	0.4716
Error	27.5400		

The ANOVA Procedure

Level of SEGMENT	N	VALUE	
		Mean	Std Dev
1	13	83.0000000	34.1345377
2	22	92.0909091	37.9560010

Conclusion: You cannot reject the null hypothesis that mean 1 = mean 2 at 95 percent confidence (this is the default value).

■ **Note** The assumption of equal variances (homogeneity of variances) can be checked with the hovtest option where you have used Levene's test. Here the p-value is .5455, and therefore you cannot reject the null.

The Welch test is for testing the hypothesis assuming no homogeneity, and the p-value for this is .4716. Therefore, you cannot accept the null.

Here's a problem for you to practice: do this for the other variables in the data set and make a conclusion about all five variables together. By doing this, you will need to create a data set with all five variables (X1 through X5).

Case Study Using R

A magazine does a survey and wants to publish the results. The following data represents business startup costs (in thousands of dollars) for shops, gathered through a survey. One of the questions to be answered is, is it equally expensive to start any of the following businesses or are some businesses cheaper to start?

```
Description of variables:
X1 = startup costs for pizza
X2 = startup costs for baker/donuts
X3 = startup costs for shoe stores
X4 = startup costs for gift shops
X5 = startup costs for pet stores
What is the conclusion that the article should publish?
```

You can start by dividing the process as per DCOVA and I.
Defining the business y is the first step.

```
# D - Define the business Y - check if the distribution of startup costs are similar across
lines of business
```

Let's brings the data into the system.

```
# Import the data
startupcost <- read.csv("H:/springer book/Case study/CaseStudy5/startupcost.csv",
stringsAsFactors=FALSE)
View(startupcost)
```

Let's look at the next two stages of collect and organize.

```
# C and O - not required for this project
```

Let's now look at the fourth stage of visualize.

```
# V - Visualise the data to establish if the data is parametric or non parametric . This
will influence the tests.
```

Use the str code to check the type of variables.

```
str(startupcost)
```

```
> str(startupcost)
'data.frame':   38 obs. of   5 variables:
 $ X1: int  80 125 35 58 110 140 97 50 65 79 ...
 $ X2: int  150 40 120 75 160 60 45 100 86 87 ...
 $ X3: int  48 35 95 45 75 115 42 78 65 125 ...
 $ X4: int  100 96 35 99 75 150 45 100 120 50 ...
 $ X5: int  25 80 30 35 30 28 20 75 48 20 ...
```

Make a copy of the data frame startupcost.

```
startup2 <- startupcost

recon1 <- rowSums(!is.na(startup2[-(1:5)]))
```

■ **Note**   No NA values show up in the data frame.

You know that the assumption is of normal distributions and equal variances. You need to test for these.

```
# check for EQUAL variance
```

```
> var.test(startup2$X1,startup2$X2)
```

Here is the f-test to compare two variances:

```
data:  startup2$X1 and startup2$X2
F = 0.8088, num df = 12, denom df = 21, p-value = 0.721
alternative hypothesis: true ratio of variances is not equal to 1
95 percent confidence interval:
 0.306730 2.472794
```

```
sample estimates:
ratio of variances
        0.8087739
```

■ **Note**   You cannot reject the null because the p-value is greater than .05. Hence, you have to accept the null hypothesis that the variance is equal.

Here's a problem for you to practice: do this for the other variables in the data frame and make a conclusion on the variance.

Check for the normal distribution of the data variables.

```
# Normal distribution
```

```
d <- density(na.omit(startup2$X1))
plot(d)
```

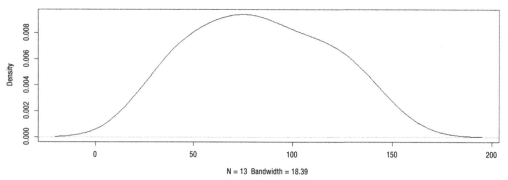

I want to use the package pastecs to work further. Since the package is already there, I am just adding the library.

```
library("pastecs")
```

Let's look at the basic statistics relating to the data.

```
stat.desc(na.omit(startup2$X1))
```

```
> stat.desc(na.omit(startup2$X1))
                     x
nbr.val       13.0000000
nbr.null       0.0000000
nbr.na         0.0000000
min           35.0000000
max          140.0000000
```

```
range           105.0000000
sum            1079.0000000
median           80.0000000
mean             83.0000000
SE.mean           9.4672174
CI.mean.0.95     20.6272947
var            1165.1666667
std.dev          34.1345377
coef.var          0.4112595
```

```
boxplot(startup2$X1)
```

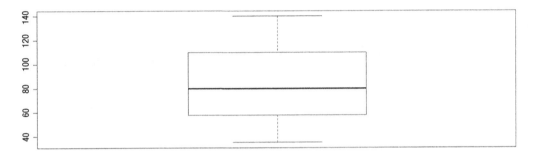

Conclusion: The mean is the median, and the boxplot shows a normal distribution.

Here's a problem for you to practice: do this for the other variables in the data frame and make a conclusion about the distribution.

Let's look at the fifth stage in the DCOVA and I process.

```
# A - Analyse . Run the t tests and ANOVA and Chi sq test
```

Let's do the t-test.

```
#T tests - Equal Variance
```

```
# t test
t.test(startup2$X1 , startup2$X2 ,var.equal = TRUE)
```

```
> t.test(startup2$X1 , startup2$X2 ,var.equal = TRUE)
```

Here are the two sample t-tests:

```
data:  startup2$X1 and startup2$X2
t = -0.7098, df = 33, p-value = 0.4828
alternative hypothesis: true difference in means is not equal to 0
95 percent confidence interval:
 -35.14897  16.96715
sample estimates:
mean of x mean of y
 83.00000  92.09091
```

Conclusion: If the p-value is less than or equal to .05, then you can reject the null. Here you will have to accept the null that the means are equal.

```
#T tests - Un-equal Variance
```

This is the default test to use. Assume unequal variance whenever you want to avoid doing the test of variance.

```
t.test(startup2$X1 , startup2$X2 ,var.equal = FALSE)
```

Here's a problem for you to practice: on the basis of the outcome of the test for equal variance, run the t-tests for unequal variance.

```
# ANOVA
```

Let's use ANOVA to check how this works.

```
str(startup2)
data1<- aov(X1 ~ X2+X3+X4, data=startup2)

summary(data1)

> summary(data1)
            Df Sum Sq Mean Sq F value Pr(>F)
X2           1   1572  1571.6   1.196  0.303
X3           1    180   180.5   0.137  0.719
X4           1    405   404.7   0.308  0.592
Residuals    9  11825  1313.9
25 observations deleted due to missingness
```

Conclusion: If the p-value is less than or equal to .05, then you can reject the null. Here you will have to accept the null that the means of the variables are equal.

- X1 = Startup costs for pizza restaurant

- X2 = Startup costs for bakery

- X3 = Startup costs for shoe store

- X4 = Startup costs for gift shop

The is similar as per the result of ANOVA.
Assumptions: Residuals should be normally distributed. This is the assumption in the statistical procedure.

```
data1$residuals

> data1$residuals
          1          2          3          4          5          6
  13.281351  31.464608 -36.297836 -28.239742  45.878150  29.793708
          7          8          9         10         11         12
  12.354359 -34.993888 -25.068705  -8.550637 -47.857625  21.052730
         13
  27.183529
```

```
shapiro.test(data1$residuals)
> shapiro.test(data1$residuals)
```

Here's the Shapiro-Wilk normality test:

```
data:  data1$residuals
W = 0.9164, p-value = 0.2242
```

Conclusion: Since the p-value is greater than .05, you cannot reject the null and have to accept the null that the data is normally distributed.

Let's now explore if the variances are equal across the four samples considered.

```
bartlett.test(list(startup2$X1,startup2$X2,startup2$X3,startup2$X4))
> bartlett.test(list(startup2$X1,startup2$X2,startup2$X3,startup2$X4))
```

Here's the Bartlett test of the homogeneity of variances:

```
data:  list(startup2$X1, startup2$X2, startup2$X3, startup2$X4)
Bartlett's K-squared = 0.9174, df = 3, p-value = 0.8212
```

Conclusion: Since the p-value is greater than .05, you cannot reject the null and have to accept the null that the data variables have equal variances.

BARTLETT'S TEST

Bartlett's test allows you to compare the variance of two or more samples to determine whether they are drawn from populations with equal variance. It is suitable for normally distributed data. The test has the null hypothesis that the variances are equal and the alternative hypothesis that they are not equal.

This test is useful for checking the assumptions of an analysis of variance.

Here's a problem for you to practice: practice ANOVA considering all five variables.

Let's now see what type of insights you can conclude about this problem.

```
# Insight generation
```

```
> summary(startup2)
      X1              X2              X3              X4
 Min.   : 35    Min.   : 40.00   Min.   : 35.0   Min.   : 35.0
 1st Qu.: 58    1st Qu.: 63.75   1st Qu.: 45.0   1st Qu.: 50.0
 Median : 80    Median : 87.00   Median : 70.0   Median : 97.5
 Mean   : 83    Mean   : 92.09   Mean   : 72.3   Mean   : 87.0
 3rd Qu.:110    3rd Qu.:115.00   3rd Qu.: 95.0   3rd Qu.:100.0
 Max.   :140    Max.   :160.00   Max.   :125.0   Max.   :150.0
 NA's   :25     NA's   :16       NA's   :18      NA's   :18
      X5
 Min.   : 20.00
 1st Qu.: 29.50
```

```
Median : 49.00
Mean   : 51.62
3rd Qu.: 75.00
Max.   :110.00
```

Here's a problem for you: practice ANOVA considering all five variables and write your insights. Here is the code for a chi-square test.

```
# Chi Sq test

Install.packages("MASS")
library(MASS)

chisq.test(startup2$X1,startup2$X2,startup2$X3,startup2$X4)

> chisq.test(startup2$X1,startup2$X2,startup2$X3,startup2$X4)
```

Here is Pearson's chi-squared test:

```
data:   startup2$X1 and startup2$X2
X-squared = 117, df = 110, p-value = 0.3061
```

Conclusion: If the p-value is less than or equal to .05, then you can reject the null. Here you will have to accept the null that the means are equal.

■ ■ ■

## CHAPTER 8

Confidence Intervals and Sanctity of Analysis Using SAS and R

In this chapter, you will look at how to use confidence intervals and error values to create a conclusion.

How Can You Determine the Statistical Outcome?

In the world of statistics, you can look at what applies to the sample and try to determine the population. You know that the sample cannot be a 100 percent replica of the population. There will be minor changes, and perhaps there are major ones too. How do you figure out that the sample statistics is applicable to the population? To answer this, you look at the *confidence interval*. Confidence intervals enable you to understand the accuracy that you can expect when you take the sample statistics and apply them to the population.

In other words, a confidence interval gives you a range of values within which you can expect the population statistics to be.

In statistics there is a term called the *margin of error*, which defines the maximum expected difference between the population parameter and the sample statistic. It is often an indicator of the random sampling error, and it is expressed as a likelihood or probability that the result from the sample is close to the value that would have been calculated if you could have calculated the statistic for the population.

The margin of error is calculated when you observe many samples instead of one sample. When you look at 50 people coming in for an interview and find that 5 people do not arrive at the correct time, you can conclude that the margin of error is $5 \div 50$, which is equal to 10 percent. Therefore, the *absolute margin of error*, which is five people, is converted to a *relative margin of error*, which is 10 percent.

Now, what is the chance that when you observe many samples of 50 people, you will find that in each sample 5 people do not come at the designated time of interview? If you find that, out of 100 samples, in 99 samples 5 people do not come in on time for an interview, you can say that with 99 percent accuracy the margin of error is 10 percent.

Why should there be any margin of error if the sample is a mirror image of the population? The answer is that there is no sample that will be a 100 percent replica of the population. But it can be very close. Thus, the margin of error can be caused because of a sampling error or because of a nonsampling error.

You already know that the chance that the sample is off the mark will decrease as the sample size increases. The more people/products that you have in your sample size, the more likely you will get a statistic that is very close to the population statistic. Thus, the margin of error in a sample is equal to 1 divided by the square root of the number of observations in the sample.

$1/SqRt(n)$

© Subhashini Sharma Tripathi 2016
S. S. Tripathi, *Learn Business Analytics in Six Steps Using SAS and R*, DOI 10.1007/978-1-4842-1001-7_8

What are the other errors that you can have in a sample? You could have coverage errors, measurement errors, or nonresponse errors.

- A *coverage error* is when you are unable to access a certain part of the population.

- A *measurement error* is a bias that is building during data collection, such as the timing of the response options, and so on.

- A *nonresponse error* results from not being able to elicit some response from the population that had to be surveyed. For example, some people may be hostile or not agree to the survey.

Thus, often the sampling error is the most likely to be measured and fixed.

One way to express a sampling error is the margin of error, which is the measure of the precision of the sample estimate of the population statistic. Thus, it uses probability to check the precision of the statistic and is generally measured in a 95 percent confidence interval. This means that if more than 1,000 samples were taken to measure the height of Indian men, the average height for 999 samples would fall into the range given by the margin of error from the first sample. Therefore, the confidence interval tells you how many times you can expect the statistic to fall within the margin of error if you were to take 100 samples. It is therefore measured in percentages.

Here is a summary of what you have learned so far:

- The margin of error is the range in which the population statistic can be expected to be in after you have measured the statistic for the sample. Thus, if you find that the average height of an Indian man in the sample is 5'6" and the margin of error is 6 inches, you can say that the population statistic of the mean height of Indian men is between 5 feet and 6 feet.

- The confidence interval tells you how many times out of 100 samples you can expect that the sample statistic will be in the range described earlier, which is 5 feet to 6 feet.

How well the sample statistic estimates the population value is addressed by the confidence interval, which provides a range of values that are likely to contain the population statistic under consideration.

Conversely, you can choose a confidence level, say, 95 percent (the default for most statistical procedures) and then calculate the sample statistic.

Therefore, you will be able to see with 95 percent confidence that in 95 out of 100 cases/samples the sample statistic would lie in the range specified, which is plus or minus 5 percent of the population statistic.

Generally, the confidence limits are seen with respect to the mean value of the population. Since the mean value of 100 samples will not all be the same and will follow a normal distribution (as per the central limit theorem studied in the previous chapter), the confidence interval estimates the lower and upper limits of the mean. The narrower the interval, the more precise the estimate and the more chance that the sample mean is equal to the population mean.

This concept is used for inferencing and concluding on the number at hand. If the hypothesis is that the mean value of sample 1 is equal to the mean value of sample 2 (Mean 1 = Mean 2) and you want to check this hypothesis at 95 percent confidence, you are trying to see that if 100 times samples were taken and compared, can you be assured in 95 of the times (out of 100 comparisons) that Mean 1 = Mean 2.

You can reject the null hypothesis that Mean 1 = Mean 2 if the test statistics (Mean 2) are outside the critical value. The critical values are the range within which 95 percent of the values lie.

You fail to reject the null if the test statistics (Mean 2) are within the critical value. The critical values are the range within which 95 percent of the values lie.

Thus, at 95 percent confidence you *reject* or *fail to reject the null.*

The z-score is used to arrive at the standard normal distribution. Thus, in a z-distribution (as per the empirical rule), the following is true:

- 68 percent of the values lie between -1 to +1 z-score.

- 95 percent of the values lie between -2 to +2 z-score.

- 99.7 percent of the values lie between -3 to +3 z-score.

z-Values for Selected (Percentage) Confidence Levels	
Percentage Confidence	**z-Value**
80%	is -1.28 to + 1.28
90%	is -1.645 to +1.645
95%	is 1.96 to +1.96
98%	is -2.33 to +2.33
99%	is -2.58 to +2.58

Thus, if you know the z-score values and the standard deviation, you can calculate the critical values within which the test statistics need to lie so as to be accepted at the given confidence percentage.

z-Values for Selected (Percentage)		Critical Values - Range	
Percentage Confidence	**z-Value**	**Critical value (lower)**	**Critical value (upper)**
80%	is -1.28 to + 1.28	Mean-1.28*SD	Mean+1.28*SD
90%	is -1.645 to +1.645	Mean-1.645*SD	Mean+1.645*SD
95%	is 1.96 to +1.96	Mean-1.96*SD	Mean+1.96*SD
98%	is -2.33 to +2.33	Mean-2.33*SD	Mean+2.33*SD
99%	is -2.58 to +2.58	Mean-2.58*SD	Mean+2.58*SD

You can do your inference: if two samples have means that can be said to be equal at the said confidence percentage.

The confidence interval defines the percentage of the sample in which the true value will exist. This means that if the confidence level is set at 99 percent, the true value will exist in 99 out of 100 samples. The confidence limits reflect a result at the level of the groups of measurement.

The size of the confidence interval depends on the size of the sample, and if the standard deviation of the sample size is large, this leads to more confidence. Conversely, if the sample is small, the confidence interval will be wide. If the dispersion or spread of values is high, then the confidence interval is wider. For the same population and 99 percent confidence, the interval is wider than a 95 percent confidence interval.

What Is the P-value?

You know that hypothesis testing is used to confirm or reject whether two samples belong to the same population. That *p-value* is the probability that determines whether the two samples are the same population. This probability is a measure of evidence *against* the hypothesis.

Remember the following:

- The null hypothesis always claims that Mean 1 = Mean 2.

- The aim of hypothesis testing is to reject the null.

Thus, a smaller p-value will mean that you can reject the null because the probability of the two samples having similar means (which points to the two samples coming from the same population) is much less (.05 = 5 percent probability).

The small p-value corresponds to strong evidence, and if the p-value is below a predefined limit (.05 is the default value in most software), then the result is said to be statistically significant.

For example, if the hypothesis is that a new type of medicine is better than the old version, then the first attempt is to prove that the drugs are not similar (that any similarity is so small that it can be random/coincidence). Then the null hypothesis of the two drugs being the same needs to be rejected. A small p-value signifies that the probability of the null hypothesis being true is so small that it can be thought to be purely by chance (see Figure 8-1).

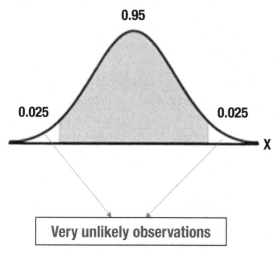

P value is the probability of the result occurring because of coincidence / chance

Figure 8-1. *Unlikely observations*

This distribution is the distribution of the probability of the null hypothesis being true. Thus, when the p-value (the probability of the null hypothesis of being true) is less than .05 (or any other value set for the test), you have to reject the null and conclude that Mean 1 = Mean 2 only because of coincidence or fate or chance.

Errors in Hypothesis Testing

No hypothesis test is 100 percent certain. As you have noticed, tests are based on probability, and therefore, there is always a chance of an incorrect conclusion. These incorrect conclusions can be of two types (see Figure 8-2).

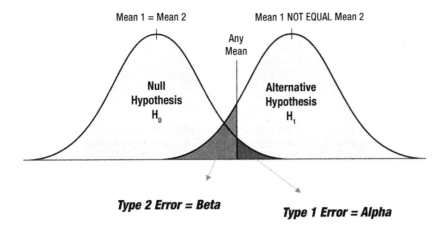

Figure 8-2. *The two types of error*

- *Type 1 error, alpha*: This is when the null hypothesis is true but you reject the null. Alpha is the level of significance that you have set for the test. At a significance of .05, you are willing to accept a 5 percent chance that you will incorrectly reject the null hypothesis. To lower the risk, you can choose a lower value of significance. A type I error is generally reported as the p-value.

- *Type II error, beta*: This is the error of incorrectly accepting the null. The probability of making a type II error depends on the power of the test.

You can decrease your risk of committing a type II error by ensuring your sample size is large enough to detect a practical difference when one truly exists.

The confidence level is equivalent to 1, the alpha level. When the significance level is 0.05, the corresponding confidence level is 95 percent.

- If the p-value is less than the significance (alpha) level, the hypothesis test is statistically significant.

- If the confidence interval does not contain the null hypothesis value between the upper and lower confidence limits, the results are statistically significant (the null can be rejected).

- If the p-value is less than the alpha, the confidence interval will not contain the null hypothesis value.

    1. *Confidence level + alpha = 1.*

    2. *If the p-value is low, the null must go.*

    3. *The confidence interval and p-value will always lead to the same conclusion.*

The most valuable usage of hypothesis testing is in interpreting the robustness of other statistics generated while solving the problem/doing the project.

- *Correlation coefficient*: If the p-value is less than or equal to .05, you can conclude that the correlation is actually equal to the correlation coefficient value displayed/calculated. If the p-value is greater than .05, you have to conclude that the correlation is because of chance/coincidence.

- *Linear regression coefficents*: If the p-value is less than or equal to .05, you can conclude that the coefficients are actually equal to the value displayed/calculated. If the p-value is greater than .05, you have to conclude that the coefficients are because of chance/coincidence.

Case Study in SAS

The U.S. government wants to understand the net inflow and outflow of cash from the country every year. U.S. trade in goods and services is summarized in terms of balance of payments. The balance of payments is the difference between exports and imports and signifies whether there was a net outflow of cash from the United States or a net inflow. This data has been provided from the U.S. Census Bureau's Economic Indicator Division.

This data is available for goods and services for the years 1960 to 2015.

■ **Note**   All values are in millions of dollars.

Questions:

- Is the balance of payments of goods and services correlated?

- Is the correlation real or coincidence?

Let's start with the define stage in the DCOVA and I process. You want to understand the correlation between the balance of payments of goods and services. You start by getting the data into SAS.

```
/* IMPORT DATA*/
PROC IMPORT DATAFILE='/saswork/SAS_work7F170001C26A_odaws01-prod-sg/#LN00010'
DBMS=CSV
OUT=WORK.BOP;
GETNAMES=YES; RUN ;
```

Let's check how the data looks.

```
/* CHECK THE DIMENSIONS OF DATA*/
PROC CONTENTS DATA=WORK.BOP; RUN ;
```

#	Variable	Type	Len	Format	Informat
3	Goods_BOP	Num	8	BEST7.	BEST7.
6	Goods_Exports	Num	8	BEST7.	BEST7.
9	Goods_Imports	Num	8	BEST7.	BEST7.
4	Services_BOP	Num	8	BEST6.	BEST6.
7	Services_Exports	Num	8	BEST6.	BEST6.
10	Services_Imports	Num	8	BEST6.	BEST6.
5	Total_Exports	Num	8	BEST7.	BEST7.
8	Total_Imports	Num	8	BEST7.	BEST7.
1	Year	Num	8	BEST4.	BEST4.
2	total_BOP	Num	8	BEST7.	BEST7.

Alphabetic List of Variables and Attributes

Conclusion: You see that all the variables are numeric.
Next you want to explore basic statistics about the data.

```
/* CHECK DESCRIPTIVE STATS*/

PROC MEANS DATA=WORK.BOP; RUN ;
```

Page Break
The MEANS Procedure

Variable	N	Mean	Std Dev	Minimum	Maximum
Year	56	1987.50	16.3095064	1960.00	2015.00
total_BOP	56	-189058.59	239585.65	-761716.00	12404.00
Goods_BOP	56	-239362.64	291675.03	-837289.00	8903.00
Services_BOP	56	50304.00	65629.05	-1384.00	233138.00
Total_Exports	56	687748.64	703404.32	25940.00	2343205.00
Goods_Exports	56	490980.84	489177.70	19650.00	1632639.00
Services_Exports	56	196767.80	214803.49	6290.00	710565.00
Total_Imports	56	876807.23	921079.29	22208.00	2851529.00
Goods_Imports	56	730343.43	769980.37	14537.00	2374101.00
Services_Imports	56	146463.80	151792.07	7671.00	490613.00

Conclusion: You understand that there are no null or NA values.
There is no need for the collect stage since you do not need to add any data. In the organize stage, you can see that there are no missing values. Now you should check for outliers.

```
/* Check Outliers */

PROC UNIVARIATE DATA=WORK.BOP;
VAR Total_Exports Total_Imports total_BOP;
RUN ;
```

You should define what you want to consider as outliers for the data variables.

```
/* limits for outliers
Lower limit = Q1 -(1.5*(Q3-Q1))
Upper limit = Q3 + (1.5*(Q3-Q1))*/

DATA WORK.BOP3;
SET WORK.BOP ;
WHERE Total_Exports  BETWEEN 2373485  AND -1254379
AND  Total_Imports BETWEEN 3399729.5 AND -1872218.5; RUN ;
```

You run the contents procedure to understand the new data set better, after deleting the outliers.

```
PROC CONTENTS DATA=WORK.BOP3; RUN ;
```

The CONTENTS Procedure

Data Set Name	WORK.BOP3		Observations	56
Member Type	DATA		Variables	10
Engine	V9		Indexes	0
Created	05/07/2016 20:53:43		Observation Length	80
Last Modified	05/07/2016 20:53:43		Deleted Observations	0
Protection			Compressed	NO
Data Set Type			Sorted	NO
Label				
Data Representation	SOLARIS_X86_64, LINUX_X86_64, ALPHA_TRU64, LINUX_IA64			
Encoding	utf-8 Unicode (UTF-8)			

Engine/Host Dependent Information

ɪt Page Size 131072

Now let's do the analyze phase of DCOVA and I.

You want to analyze the data by running a correlation between the balance of payments of goods and services.

```
/* A - CORRELATION */

PROC CORR DATA=WORK.BOP3;
VAR Total_Exports Total_Imports; run ;
```

Page Break
The CORR Procedure

2 Variables: Total_Exports Total_Imports

Simple Statistics						
Variable	N	Mean	Std Dev	Sum	Minimum	Maximum
Total_Exports	56	687749	703404	38513924	25940	2343205
Total_Imports	56	876807	921079	49101205	22208	2851529

Pearson Correlation Coefficients, N = 56 Prob > |r| under H0: Rho=0		
	Total_Exports	Total_Imports
Total_Exports	1.00000	0.99227
		<.0001
Total_Imports	0.99227	1.00000
	<.0001	

Conclusion: The correlation coefficient is .99, and the p-value is less than or equal to .05. Thus, you can conclude that the correlation is real and should be accepted.

As an exercise, you should run the correlation for the following:

- Goods_Exports vs. Services_Exports

- Goods_Imports vs. Services_Imports

What is your conclusion?

Case Study with R

The U.S. government wants to understand the net inflow and outflow of cash from the country every year. U.S. trade in goods and services is summarized in terms of a balance of payments. The balance of payments is the difference between exports and imports and signifies whether there was a net outflow of cash from the United States or a net inflow. This data has been provided from the U.S. Census Bureau's Economic Indicator Division.

This data is available for goods and services for the years 1960 to 2015.

■ **Note**   All values are in millions of dollars.

Questions:

- Is the balance of payments of goods and services correlated?

- Is the correlation real or coincidence?

Let's start with the define phase of the DCOVA and I process. You want to understand the correlation between the balance of payments of goods and services.

You start the code part by bringing the data into the R tool.

```
# Import data

BOP <- read.csv("F:/springer book/Case study/CaseStudy6/BOP.csv", stringsAsFactors=FALSE)
```

Let's see the format of the variables by looking at the str code.

```
# Check Format
str(BOP)
```

Conclusion: All the data variables are integers.

Now you want to check some of the basic statistics about the data to get a sense of how the data looks.

```
# Check Descriptive statistics
```

You will work with the package pastecs, which has a nice descriptive statistics output.

```
install.packages("pastecs")

library(pastecs)

stat.desc(BOP)
```

Conclusion: There are no null or NA values.

Let's look at the data in a box plot. You will be able to check for outliers too.

```
# Draw boxplot

boxplot(BOP)
```

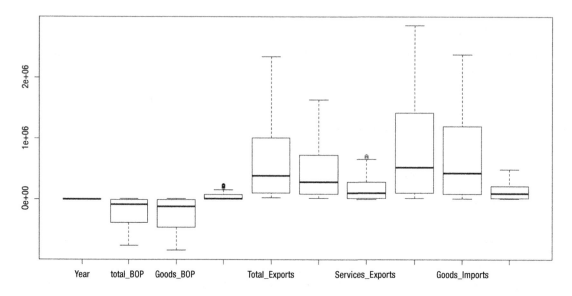

Conclusion: There are outliers in most of the variables.

You need to remove the outliers for total_BOP, Goods_BOP, and Services_BOP.

Let's remove the outliers using the boxplot.stats function; it is an ancillary function that produces statistics for drawing box plots. It returns, among other information, a vector stat with five elements: the extreme of the lower whisker, the lower "hinge," the median, the upper "hinge," and the extreme of the upper whisker. The extremes of the whiskers are the adjacent values (the last nonmissing value; that is, every value beyond is an outlier).

The outliers are then all values outside the interval id1$stats[1] < and id1$stats[5].

```
id1 <- boxplot.stats(BOP$total_BOP)

id2 <- boxplot.stats(BOP$total_BOP, coef=2)

id1$stats
> id1$stats
[1] -761716.0 -378145.5  -87002.5   -2797.5   12404.0

id1$stats[1]
> id1$stats[1]
[1] -761716

id1$stats[5]
> id1$stats[5]
[1] 12404
```

Conclusion: The outliers for BOP$total_BOP is less than -761716 and greater than 12404.

```
summary(BOP$total_BOP)
> summary(BOP$total_BOP)
   Min. 1st Qu.  Median    Mean 3rd Qu.    Max.
-761700 -375300  -87000 -189100   -3545   12400
```

Conclusion: There are no outliers for total_BOP.

```
id1 <- boxplot.stats(BOP$Goods_BOP)
id2 <- boxplot.stats(BOP$Goods_BOP, coef=2)
id1$stats
id1$stats[1]
id1$stats[5]
summary(BOP$Goods_BOP)
```

Conclusion: There are values below the lower limit of -837289 that should be dropped.

```
id1 <- boxplot.stats(BOP$Services_BOP)
id2 <- boxplot.stats(BOP$Services_BOP, coef=2)
id1$stats
id1$stats[1]
id1$stats[5]
summary(BOP$Services_BOP)
```

Conclusion: There are values greater than 154020 that should be dropped.

```
# We will subset to remove outliers from 2 variables
NEWBOP<- subset(BOP, Goods_BOP>-837289)
NEWBOP<-subset(NEWBOP, Services_BOP<154020)
dim(NEWBOP)
```

check the new subset data using 'dim' code.

```
> dim(NEWBOP)
[1] 49 10
```

Now let's look at the analysis in the DCOVA and I process.

```
# A- Analyse the data by running correlation between Services and Good BOP

cor(NEWBOP$Goods_BOP, NEWBOP$Services_BOP)
cor.test(NEWBOP$Goods_BOP, NEWBOP$Services_BOP)$p.value

> cor(NEWBOP$Goods_BOP, NEWBOP$Services_BOP)
[1] -0.7944912
> cor.test(NEWBOP$Goods_BOP, NEWBOP$Services_BOP)$p.value
[1] 9.497043e-12
```

Conclusion: The correlation coefficient is -.79, and the p-value is less than or equal to .05. Thus, you can conclude that the correlation is real and should be accepted.

Note You can also use the correlation function in the Hmisc package.

```
#install.packages("Hmisc")
#library(Hmisc)
# rcorr(NEWBOP, type="pearson")
```

As an exercise, you should run the correlation for the following:

- a. Goods_Exports vs. Services_Exports

- b. Goods_Imports vs. Services_Imports

What is your conclusion?

CHAPTER 9

Insight Generation

In this chapter, you will learn about the types of conclusions that can be drawn for an analytics project and how you can interpret the results generated by SAS and R.

Introducing Insight Generation

What happens when there is a mystery? In other words, what happens when the management in a business enterprise finds something that is not a satisfactory at the moment? They will present it as a problem or project to the analytics team. You and your team have to solve the mystery, investigate to find out what is happening and why, look for clues, and solve the problem.

There are many business mysteries. For example, who is the most profitable customer? Which type of customer should you avoid? Which vendor is the best vendor for the business? What type of problems do your customers generally approach you for? How can you ensure that you get a heads up before a customer leaves? Which employee is most suited for a particular type of job? What are the costs that can be most easily reduced?

As a good investigator knows, there is virtually no case in which there is 100 percent certainty of solving the case until the final step in the case. Thus, detectives and investigators generally use logic where the conclusions are more probable. Hence, they have to explore multiple processes and possibilities. This is quite similar to the scientific method, where a hypothesis is tested against the evidence provided by the data. Conclusions are drawn, and then a particular path is taken. If the case does not get solved, you have to start all over again with another hypothesis.

Thus, logical thinking, along with mathematical techniques, is the way that analytics resources solve a particular problem. This is also the reason that sometimes they are unable to solve the problem if the data gives solutions where the probability of occurrence is lower than the threshold limit or your fault tolerance limit. You cannot run the risk of a Wrong judgment in, say, 20 percent of the cases. However, you may be OK with the wrong judgment in 5 percent of the cases.

Even for two cases with the same problem statement, the data will be different. This is indeed a confusing situation. The analytics resource will find that it pays to come up with the following:

- *Analogies*: The analyst can look at similar problems and study the solutions to get a list of hypothesis that may work for the problem at hand.

- *A process (DCOVA and I)*: This will give the analyst confidence that nothing is missing, including critical aspects of the data or critical parts of problem-solving.

Analogies help you to establish points of similarities and differences between other projects and this project. Thus, they reduce substantially the time an analyst has to spend trying to figure out possible and probable solutions.

© Subhashini Sharma Tripathi 2016
S. S. Tripathi, *Learn Business Analytics in Six Steps Using SAS and R*, DOI 10.1007/978-1-4842-1001-7_9

With statistics, you can calculate a particular thing in multiple ways, and you can solve a particular problem using different statistical methods. As you know, there are multiple methods grouped broadly under these five segments:

- Descriptive statistics

- Inferential statistics

- Differences statistics

- Associative statistics

- Predictive statistics

For a data analyst or data scientist, figuring out which statistical method is most appropriate is in itself a challenge. The DCOVA & I process gives the data scientist or analyst a way to explore and understand the data better (DCOV) so that the best statistical technique can be identified to solve the problem in the analyze phase (A).

The insight generation (I) ensures that the mathematical output is understood in simple language so that the business can create a strategy and the technology team can implement it in their systems.

Since an analytics project closely resembles a mystery/crime investigation, the results at each stage have to be understood for you to take the next step in the project. The following are the common inferences for each type of statistical procedure covered in this book.

Descriptive Statistics

Here are some notes on descriptive statistics:

- Since the mean is nearly equal to the median and the range is nearly equal to standard deviation (6SD), you can conclude that the data seems normally distributed and does not have outliers. This is validated with a histogram or density plot for the variable (or variables).

- Since the data is categorical and the number of categories is independent in meaning, the mode is the best way to track the most popular category. (Category examples are customer satisfaction rating, product variant, and so on.)

- Since the data median is less than the mean and the data is normally distributed, the data is skewed toward the smaller denominations, such as the minimum value of the range.

- Since the median is higher than the mean and the data is normally distributed, the data is skewed toward the higher denominations, such as the maximum value of the range.

- Since the sample standard deviation is greater than the population standard deviation for the same data set, the sample seems like a good sample to conduct the test/project. (This is because the formula for the sample standard deviation has to take into account the possibility of there being more variation in the true population than has been measured in the sample.)

Graphs

Here are some useful inferences in graphs:

- Pie charts:
    - The largest contributor to the said problem are Segment X followed by Segment Y for Variable 1.
    - The least important component of said problem is Segment Z for Variable 1.
- Bar graph:
    - The number of Variable 1 contributors is fairly constant across the different segments.
    - The value of the contribution of Variable 1 remains fairly stable across time.
    - There is an upward trend in Variable 1 over time.
- Line chart:
    - The line for the Variable 1 shows an increasing trend over time.
    - The lines for Variable 1 and Variable 2 move in a similar direction and show a positive correlation over time.
- Scatter plot:
    - Variable 1 has n distinct clusters emerging in the scatter plot.
    - Variable 1 and Variable 2 show positive correlation.
    - Variable 1 shows a definite trend and a normal distribution (where the scatter plot is denser in the middle and tapers off toward the end). Thus, you can run a linear regression model.

Inferential Statistics

Here are some notes on this topic:

- Normal distribution:
    - Since Variable 1 is normally distributed, as per the empirical rule, 95 percent of the values lie between Mean 1 and +_2 SD for Variable 1.
    - Since Variable 1 is not normally distributed, as per Chebyshev's theorem, at least 75 percent of the values lie between Mean 1 and +_2 SD for Variable 1.
- Probability:
    - The probability of someone doing Action 2 subsequent to doing Action 1 is x percent.
    - The probability of Events 1 and 2 occurring together is x percent.
- Sampling:
    - The sample is a good representative sample because the mean of the sample and population are the same at 95 percent confidence.

Differences Statistics

Here are some notes on differences statistics:

- Hypothesis testing:

    - The two samples are similar and can be said to belong to the same population . The p-value for the test is less than .05.

    - At least one sample in this group of samples is different. Any similarity is coincidence since the p-value is greater than .05.

- Confidence intervals:

    - The correlation value displayed is .0x and is a valid value. The p-value for the statistics is less than .05.

    - At 95 percent confidence you can state that the said statistic is true. Thus, the model is a good one.

    - The descriptive statistics for the sample will apply to the population at 95 percent confidence. Thus, the range of likely values for mean/median/SD lies between the range x,y,z.

Let's understand how these flow by looking at a case study.

Case Study with SAS

A retail company has an application to track sales named Sales. However, there are some returns in the file named Sales_returns. The company has sales managers who handle different regions in the file Sales_manager.

The company wants to understand the following:

- What are the net sales after returns have been factored in?

- What percentage of the total sales do the returns form?

- Are the returns linked to the value of items (cost)?

Let's look at the define stage of the DCOVA and I process. In this project, the define is clearly mentioned in the business problem statement. No further mathematical understanding needs to be done.

- What are the net sales after returns have been factored in?

- What percentage of the total sales do the returns form?

- Are the returns linked to value of items (cost)?

Let's start by importing the data into the SAS tool.

```
PROC IMPORT DATAFILE='/home/subhashini1/my_content/Sales.csv'
    DBMS=CSV
    OUT=WORK.SALES;
RUN;
```

You should look at the data to understand it more.

```
/* EXPLORE THE DATA */

PROC CONTENTS DATA=WORK.SALES; RUN ;
```

Now it's time for the collect stage. You get the data together for returns and sales. There is not much to do in the organize phase.

```
/* C & O - GET THE RETURNS DATA TOGETHER WITH THE SALES DATA*/


PROC IMPORT DATAFILE='/home/subhashini1/my_content/Sales_returns.csv'
    DBMS=CSV
    OUT=WORK.RETURNS;
RUN;

PROC CONTENTS DATA=WORK.RETURNS; RUN ;
```

You must always remember to sort the data sets by the primary key before merging them.

```
/* SORT THE DATASETS ON PRIMARY KEY */


PROC SORT DATA=WORK.SALES;
BY "Order ID"N; RUN ;
PROC SORT DATA=WORK.RETURNS;
BY "Order ID"N; RUN ;

DATA WORK.TOTAL;
MERGE WORK.SALES WORK.RETURNS;
BY "Order ID"N; RUN ;
PROC PRINT DATA=WORK.TOTAL (OBS=10) ; RUN ;
```

Let's look at some preliminary visualizations in tables and graphs.

```
/* V - RUN FREQ TABLE TO UNDERSTAND DATA SPLIT IN 'STATUS' BETWEEN
RETURN AND NOTRETURN*/

PROC FREQ DATA=WORK.TOTAL ;
TABLES STATUS ; RUN ;
```

The FREQ Procedure

Status	Frequency	Percent	Cumulative Frequency	Cumulative Percent
Returned	872	100.00	872	100.00
Frequency Missing = 7527				

```
DATA WORK.TOTAL;
SET WORK.TOTAL ;
IF STATUS = 'Returned' THEN STATUS2= 'RETURNED';
ELSE STATUS2 = 'NOTRETURNED'; RUN ;

PROC FREQ DATA=WORK.TOTAL ;
TABLES STATUS2 ; RUN ;
```

The FREQ Procedure

STATUS2	Frequency	Percent	Cumulative Frequency	Cumulative Percent
NOTRETURNED	7527	89.62	7527	89.62
RETURNED	872	10.38	8399	100.00

Use the button-driven menu to execute the pie chart.

You need to press the buttons in the following sequence: Tasks ➤ Graphs ➤ Pie Chart.

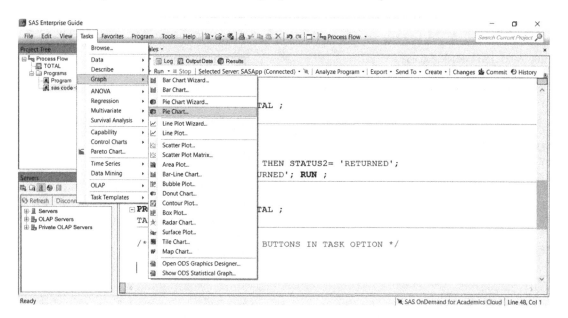

Select the data.

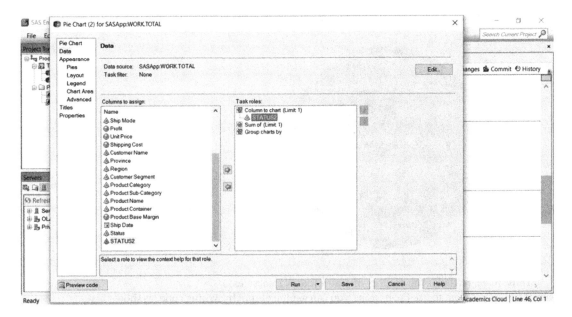

Select the layout.

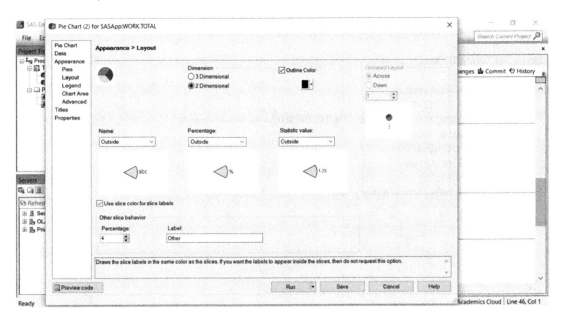

Pie Chart

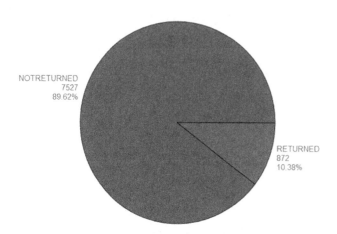

Thus, 10.38 percent is the number of returns out of total sales.

If you do not want to use the buttons, you can get the output by writing code, as shown here:

```
/* ALTERNATE METHOD - RUN PIE CHART USING CODE*/

 /* Set the graphics environment */
goptions reset=all cback=white border htitle=12pt htext=10pt;

title1 "RETURNS IN SALES";

proc gchart data=work.total;
   pie Status2 / other=0
               midpoints="RETURNED" "NOTRETURNED"
               value=none
               percent=arrow
               slice=arrow
               noheading
               plabel=(font='Albany AMT/bold' h=1.3 color=depk);
run;
quit;
```

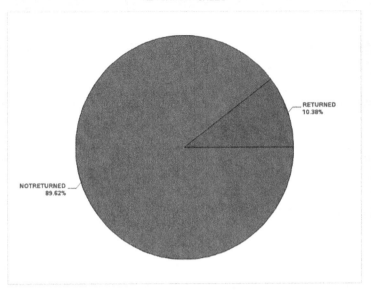

```
/* CHECK BY VALUE OF SALES*/

/* Set the graphics environment */
goptions reset=all cback=white border htitle=12pt htext=10pt;

title1 "RETURNS IN SALES";

proc gchart data=work.total;
   pie Status2 / SUMVAR=SALES
              midpoints="RETURNED" "NOTRETURNED"
              value=none
              percent=arrow
              slice=arrow
              noheading
              plabel=(font='Albany AMT/bold' h=1.3 color=depk);
run;
quit;
```

RETURNS IN SALES

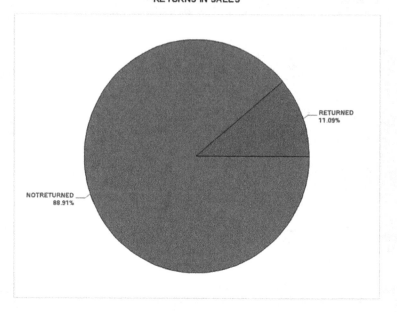

Thus, there are returns totaling 11.09 percent of the sales values.

Now let's look at the analytics part. You want to check for preliminary associations.

```
/*A - CORRELATION */

DATA WORK.TOTAL ;
SET WORK.TOTAL ;
IF STATUS2 = 'RETURNED' THEN STATUS3 = 1;
ELSE STATUS2 = 0 ; RUN ;

PROC CORR DATA=WORK.TOTAL;
VAR SALES STATUS3; RUN ;
```

RETURNS IN SALES

The CORR Procedure

2 Variables: Sales STATUS3

Simple Statistics						
Variable	N	Mean	Std Dev	Sum	Minimum	Maximum
Sales	8399	1776	3585	14915601	2.24000	89061
STATUS3	8399	0.10382	0.30505	872.00000	0	1.00000

Pearson Correlation Coefficients, N = 8399
Prob > |r| under H0: Rho=0

	Sales	STATUS3
Sales	1.00000	0.01157 0.2889
STATUS3	0.01157 0.2889	1.00000

Here are the insights:

- 10 percent of the total number of goods sold is returned.

- 11 percent of the goods is returned by value.

The correlation between the higher sales value of goods and returns is very low (.011). Since the p-value of the correlation is high (.22), you cannot conclude that this value of correlation is not by coincidence. Therefore, all evidence points to no correlation between value of goods and returns.

With this, you have come to the end of learning about the basics of analytics and using the tools SAS and R to solve problems. By now in this book, you should be feeling confident about planning for a project using the DCOVA and I methodology and executing it.

A subsequent book will cover the analytical techniques, including predictive models and clustering, using this same six-step sequence of DCOVA and I.

If you have any doubts or confusion, you can reach me at subhashini@pexitics.com or subhashini@indiadecisionmanagement.com.

Case Study in R

A retail company has an application to track sales named Sales. However, there are some returns in the file named Sales_returns. The company has sales managers who handle different regions in the file Sales_manager.

The company wants to understand the following:

- What are the net sales after returns have been factored in?

- What percentage of the total sales do the returns form?

- Are the returns linked to value of items (cost)?

Let's look at the define stage of the DCOVA and I process. In this project, the define stage is clearly mentioned in the business problem statement. No further mathematical understanding needs to be done.

- What are the net sales after returns have been factored in?

- What percentage of the total sales do the returns form?

- Are the returns linked to the value of items (cost)?

You start with bringing the data into the R tool.

```
# IMPORT DATA
Sales <- read.csv("F:/springer book/Case study/CaseStudy7/Sales.csv",
stringsAsFactors=FALSE)
```

Let's look at the data variables and how they look by using the str or dim code.

```
str(Sales)
```

```
> str(Sales)
'data.frame':            8399 obs. of  21 variables:
 $ Row.ID         : int  1 49 50 80 85 86 97 98 103 107 ...
 $ Order.ID       : int  3 293 293 483 515 515 613 613 643 678 ...
 $ Order.Date     : chr  "10/13/2010" "10/1/2012" "10/1/2012" "7/10/2011" ...
 $ Order.Priority : chr  "Low" "High" "High" "High" ...
 $ Order.Quantity : int  6 49 27 30 19 21 12 22 21 44 ...
```

```
dim(Sales)
> dim(Sales)
[1] 8399    21
```

Now let's go to the collect and organize phases.

In the collect phase, you will need to merge the returns data with the manager names.

There is nothing much to be done in the organize stage.

```
# C and O
# Merge the Returns data and the Manager name
# import the 2 datasets

Manager <- read.csv("F:/springer book/Case study/CaseStudy7/Sales_manager.csv",
stringsAsFactors=FALSE)
str(Manager)

Returns <- read.csv("F:/springer book/Case study/CaseStudy7/Sales_returns.csv",
stringsAsFactors=FALSE)
str(Returns)
```

You can sort the data before merging it as good practice.

```
# Sort data
attach(Returns)
Returns1 <- Returns[order(Order.ID),]

attach(Sales)
Sales1 <- Sales[order(Order.ID),]
```

Create a new data set after the merger called total.

```
total <- merge(Sales1,Returns1,by="Order.ID",all = TRUE)

dim(total)
> dim(total)
[1] 8399    22
```

Let's do some preliminary visualizations to get some sense of the data.

```
# Visualise - Sales vs Returns (Status = Returned / NA)
total[["Status"]][is.na(total[["Status"]])] <- "NotReturned"
```

I use the package MASS for creating cross-tabulated tables and graphs.

```
# frequency of returns
library(MASS)

mytable<- xtabs(~Status, data = total)

mytable
```

After creating the frequency table mytable, you want to view the output as a pie chart.

```
pie(mytable)

slices<- c(7527, 872)
lbls<- c("NotReturned", "Returned")
pct <- round(slices/sum(slices)*100)
lbls <- paste(lbls, pct)
lbls <- paste(lbls,"%",sep="")
pie(slices,labels = lbls, col=rainbow(length(lbls)),
    main="Returned vs Not Returned- freq")
```

The output looks like this:

Returned vs Not Returned- freq

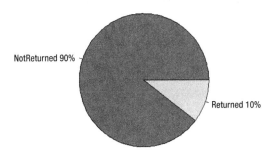

211

Let's do one more visualization on the value of the return.

```
# value of returns thru cross tab

total$Salesnum <- total$Sales
mytable2 <- aggregate(Salesnum ~ Status, total, sum)
mytable2

> mytable2
        Status Salesnum
1 NotReturned 13260747
2    Returned  1654854
```

Create the graph on the table.

```
slices<- c(13260747, 1654854)
lbls<- c("NotReturned", "Returned")
pct <- round(slices/sum(slices)*100)
lbls <- paste(lbls, pct)
lbls <- paste(lbls,"%",sep="")
pie(slices,labels = lbls, col=rainbow(length(lbls)),
    main="Returned vs Not Returned- value")
```

Returned vs Not Returned- value

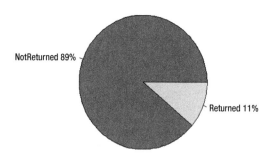

Now you are in the analyze stage. Let's understand any associations in the data. As you have studied, the correlation is

You start by changing the data to numeric data.

```
# Create numerical value for Status

total$statusnum[total$Status=="Returned"] <- 1

total$statusnum[total$Status=="NotReturned"] <- 0

cor(total$Salesnum,total$statusnum )
```

Run the correlation.

```
> cor(total$Salesnum,total$statusnum )
[1] 0.01157299
cor.test(total$Salesnum,total$statusnum )$p.value

> cor.test(total$Salesnum,total$statusnum )$p.value
[1] 0.2889189
```

Now it's time for the insights or conclusions.

```
# Insight
10% of the total count of goods sold is returned
11% of the goods is returned by value.
```

The correlation between the higher sales value of goods and returns is very low (.011). Since the p-value of the correlation is high (.22), you cannot conclude that this value of correlation is not by coincidence. Therefore, all evidence points to no correlation between the value of goods and returns.

Index

© Subhashini Sharma Tripathi 2016
S. S. Tripathi, *Learn Business Analytics in Six Steps Using SAS and R*, DOI 10.1007/978-1-4842-1001-7

Get the eBook for only $4.99!

Why limit yourself?

Now you can take the weightless companion with you wherever you go and access your content on your PC, phone, tablet, or reader.

Since you've purchased this print book, we are happy to offer you the eBook for just $4.99.

Convenient and fully searchable, the PDF version enables you to easily find and copy code—or perform examples by quickly toggling between instructions and applications.

To learn more, go to http://www.apress.com/us/shop/companion or contact support@apress.com.